The Static Universe
Exploding the Myth of Cosmic Expansion

Hilton Ratcliffe

Apeiron
Montreal

Published by C. Roy Keys Inc.
4405, rue St-Dominique
Montreal, Quebec H2W 2B2 Canada
http://redshift.vif.com

Library and Archives Canada Cataloguing in Publication

Ratcliffe, Hilton
        The static universe : exploding the myth of cosmic expansion / Hilton
Ratcliffe.

Includes bibliographical references and index.
ISBN 978-0-9864926-2-4

        1. Cosmology. I. Title.

QB982.R37 2010   523.1   C2010-900688-7

Dedication

This work is dedicated to the memory of the late Anthony Bray (1923 – 2009) and the late Tom Van Flandern (1940 – 2009). They were foremost amongst those exemplary, gentle folk who, despite holding carefully developed opinions sometimes contrary to my own, were still open enough to consider my case without hostility, and offer criticism for no other reason than to help me improve my thesis.

"There is a principle which is a bar against all information, which is proof against all arguments and which cannot fail to keep a man in everlasting ignorance—that principle is contempt prior to investigation."

Herbert Spencer

"I know that most men, including those at ease with problems of the greatest complexity, can seldom accept even the simplest and most obvious truth if it be such as would oblige them to admit the falsity of conclusions which they delighted in explaining to colleagues, which they have proudly taught to others, and which they have woven, thread by thread, into the fabric of their lives."
Leonid Tolstoy.

"What a weak barrier truth is when it stands in the way of an hypothesis."
Mary Wollstonecroft Shelley

Table of Contents

Acknowledgements

First and foremost, my work applauds the world's amateur astronomers. You are the most important people in the whole show. What you see, unadorned by theoretical shenanigans, is the bare bones of astrophysics—the optical image. My hope is that all my readers have been, or soon will be, inspired to look at the heavens.

Before this book was complete, Professor Anthony Bray passed on to the great beyond, and thus came to an Earthly end a friendship of wonderful richness. Tony, how could I ever thank you? Those endless cups of tea and Madge's cream scones, consumed while we discussed in fearless detail my efforts at capsizing the paradigm, are in my memories forever a monument to your staggering intellect and encyclopaedic knowledge. I shall try my best to continue this journey without you, but it just won't be the same.

Tony, this book is for you.

My gratitude and sincere thanks go to Sir Patrick Moore for writing the foreword. If one could line up all the people that Patrick has helped during his lifetime—and I am blessed to be one of them—they would stretch well beyond the horizon, I'm sure. I've said it before and I'll say it again: When it comes to real knowledge of real things in the cosmos, Patrick Caldwell Moore is without peer.

Huge thanks are due to Roy Keys, who undertook the publication of this work. His was a demonstration of patience and understanding the likes of which I seldom see. Dave Roscoe cheerfully undertook the miserable task of doing a final sweep of the manuscript, and his comments were invaluable. How he kept his sense of humour I'll never know!

As always, there are those kind folk of immense knowledge and experience who help me, and I can't for the life of me figure out why they so cheerfully apply their enthusiasm to honing my down-home astrophysics. It must be the intellectual equivalent of dental surgery, yet David Raal and Paul Jackson—professors of physical science both—are unstinting in their assistance and passion for real science.

Special mention must go also to Andre Assis. It was inspiring to find someone so prominent in the world of physics who shares my appreciation of classical mechanics and aversion to the blatant fabrication of evidence. John Hartnett is a special influence with a unique perspective, applied in the form of invaluable comments during the drafting of this work. He reminds me always where I came from and where I'm going, and intervenes promptly when I get too big for my boots. John B. Eichler took valuable time out from his studies and the writing of his own book to help me with mine. Pierre-Marie Robitaille was generous in his support and encouragement, and in Dr Robitaille I found synergy with someone who has for far longer than I "swum against the stream" while holding aloft the flag of reality physics.

I am extremely grateful to the owners of the diagrams illuminating the pages of this book for kindly allowing me to display them here. Special thanks go to NASA's Hubble Heritage and the Space Telescope Science Institute for freely allowing general use of the magnificent Hubble Space Telescope images. Martín López-Corredoira, as always, goes out of his way to help me, and kindly allowed me to use his wonderful images without restriction. Likewise, I salute Halton "Chip" Arp, who gave me *carte blanche* with his published material. Some of his line diagrams were scanned from the pages of his publications and are consequently a bit fuzzy, but I've left them like that because they lend an authentic charm to what might otherwise be sterile science graphics. I was also graciously given open access by Y.P. Varshni, to whom I also extend my thanks. Warm appreciation goes to Gregg Barlow for his patient draughtsmanship in several pictures, including the Klein bottle on the cover.

It is often difficult to establish just what a broad, somewhat esoteric theory like Big Bang actually consists of, and even more exasperating trying to get the latest version. They are often distilled out of a melting pot of ideas from a few radical thinkers, and subsequently mature into a form that usually represents an ill-defined consensus between like-minded scientists. No one owns the theory, and no one has the right to edit it at will. The so-called Concordance Cosmology, officially titled the Lambda-Cold Dark Matter Model, is necessarily painfully technical, and that just won't do for my anticipated readership.

With this in mind, I made a reasonably educated choice of just whom it is I am going to refer to for the principles of Big Bang Theory. I chose P.J.E. "Jim" Peebles. The pioneering works of Lemaître and Gamow were soon outdated by the rapid advance of cosmology in the Hubble era, and were in any case couched in rhetoric. Those gentlemen had other agendas, and I sought more modern, less subjective analysis to work from. Jim Peebles' masterpiece, *The Principles of Physical Cosmology* (Princeton: Princeton University Press, 1993) is ideal; the author is an icon in space science, sincerely in favour of the Standard Model of Cosmology, and backed by impeccable qualifications and experience in the field. His book should be prominent on the bookshelf of any student of modern cosmology.

I have tried to espouse the legendary Dick Feynman's dictum that theories should be explained so that ordinary folk could understand them, and to that end sought references that put Big Bang Theory in the language of a lay readership. The primary document I use for this purpose is published on an Internet forum called *The TalkOrigins Archive*. It is entitled "Evidence for the Big Bang," and the authors are Björn Feuerbacher and Ryan Scranton. The Web address is

http://www.talkorigins.org/faqs/astronomy/bigbang.html.

I have chosen this summary in preference to Ned Wright's better-known "Cosmology Tutorial"[*] because it is in my opinion more appropriately written for the purposes of this book. Thank you, Björn and Ryan, for laying it on the line for us, and especially for doing it so clearly and unambiguously. We may disagree, but I respect your well-considered opinions.

On the other side of the coin, I found Mike Disney's excellent review "The Case Against Cosmology" (arXiv: astro-ph/0009020) invaluable as a non-aligned summary of cosmology's fundamental tenets, including the principles of universal expansion and in-

---

[*] www.astro.ucla.edu/~wright/cosmolog.htm

flation. Also recommended are references to Tom Van Flandern's "Meta Research Top 30 Problems with Big Bang" (www.metaresearch.org/), and Eric Lerner's seminal book *The Big Bang Never Happened* (New York :Vintage Books, 1992). Eric's home page is www.bigbangneverhappened.org. A latecomer to my bibliography is Lyndon Ashmore's lucid frontal attack, *Big Bang Blasted* (Charleston SC: BookSurge, 2006).

Although it is at the time of writing some 5 years old, Dr Martín López-Corredoira's review paper "Observational Cosmology: caveats and open questions in the standard model" (arXiv: astro-ph/0310214 v2) is really worth looking at carefully. Please print yourself a copy and keep it to hand. These are the words of a hands-on observational astronomer with two doctorates (written in his pre Instituto de Astrofísica de Canarias days while he was still at Astronomisches Institut der Universität Basel), and that makes this reference particularly pertinent to the way I do astrophysics: Observation before theory!

Another invaluable reference source is the monthly ACG newsletter, available free and without obligation from www.cosmology.info/newsletter. You can save yourself a great deal of time by flipping through the ACG newsletter. It is a review of all papers published on arXiv in the category astro-ph for the month in question. Eric Lerner and I are the editors, and we scan more than 1,000 papers each month. I will be using it extensively in compiling this work.

It is important that I emphasise the philosophical component of my writing. Science, after all, is a set of quantified and quantumized thought-bytes leading, I would hope, to the expression of a systematic understanding of nature. Perhaps surprisingly to those not playing the game, the outcome of scientific investigation is profoundly affected by the "choice of priors," as physicists are wont to put it. The current crisis in cosmology is simply a perfectly natural consequence of philosophical preferences. Mine (give or take a few minor grey areas) are aligned with those expressed so incisively by Thomas Kuhn and Karl Popper in their books *The Structure of Scientific Revolutions* and *The Logic of Scientific Discovery*, respectively. Please refer to the bibliography at the back of this book for details.

Pat and Ann Cole-Bowen have been stalwarts. For more than thirty years they put up with me, offering warmth, encouragement, and guidance when the way was dark, and unconditional

friendship when it was light. My mother-in-law Rose van Staden
let her gentle soul whisper across my mind, filling the gaps with
Persian poetry when nothing else would do. Rose and I plait
wonderful, wistful memories of Monty and Heidi, both taken
from us before we could fully realise what they meant. Mark
Sandison, a well read philosopher, has from his first glimpse of
The Virtue of Heresy, through his subsequent line-by-line analysis
of *The Static Universe*, been indefatigable: He is convinced, and he
says so, often. Budding space scientist Chanel Henry has also
shared her infectious enthusiasm with me. I can see her name in
lights one day! I cannot forget my dear friend, outgoing President
of the Astronomical Society of Southern Africa, Magda Streicher.
We share a deep love for the heavens and observation, and her
column "Deep Sky Delights" in MNASSA really is a delight.
Thank you, Magda!

All my writing has been characterised by the music track that
fed my soul while I struggled with an avalanche of ideas and
nowhere to go. *The Virtue of Heresy* was wrought in the gentle
acoustic and semi-electric crib of Van Morrison's early albums *As-
tral Weeks* and *Moondance*. This book is driven by an altogether
different genre, a totally divergent stream of musical consciousness for which I must thank my friends Geoffrey and Shirley Sta-
pleton.

In 2007, I travelled up to Abingdon to spend a few days with
the Stapletons. In their charming, modern home just far enough
from Oxford to be rustically peaceful, evenings would see Geoffrey select some of his favourites from an impressive selection of
neatly arranged CDs to set the tranquil mood.

The music was clear and inspiring. I returned to South Africa
and acquired a collection of what I thought to be the kind of music Geoffrey had played for us (he has since pointed out that I
didn't get it quite right, but near enough). Consequently, a fairly
eclectic mix of Mozart Nachtmusik, Barbara Hendricks and Maria
Pires' renditions of his Lieder, and Frederick Chopin nocturnes at
the hands of Vitalij Margulis, creatively fuelled these and the
pages ahead.

Now, a brickbat! The following paragraph was written before
I was (quite rudely, I thought) blacklisted by arXiv. After a deal of
sombre thought, I decided to leave it here, unchanged:

It is about time that someone gave credit to the most-used
reference set in the history of science: The well-worn Cornell

University online library arXiv. Pronounced archive from the Greek letter Chi, arXiv currently stores about 500,000 scientific publications, with about 4,000 being added every month. Access is free and open, and it is the preferred point of reference for scientists seeking to refer to the work of others. What an outstanding service! Thank you so much, Cornell for administering it, and Paul Ginsparg for inventing it.

That was said in all sincerity. I'm sure you will understand that I am somewhat more cynical about arXiv these days. It presents an imbalance—the absence of even a few of those who argue against the motion means that arXiv becomes the expression of a particular opinion rather than a place where scientific results can be compared without let or hindrance. An alternative to arXiv has recently been launched by physicist Phil Gibbs: It is called viXra and can be found at www.viXra.org. I hope it will be well supported so that it can become a viable resource for science.

Finally, I blow my own trumpet. My book *The Virtue of Heresy—Confessions of a Dissident Astronomer* (Charleston SC : Book-Surge, 2008) is now in its third edition, revised and updated. It is a comprehensive introduction to the crisis in science, and a useful prelude to the current work.

Hilton Ratcliffe
Durban, South Africa
2010

Foreword

By Sir Patrick Moore, CBE, FRS

Hilton Ratcliffe is a highly-qualified professional astro-physicist, working actively on problems of cosmology; I am an amateur who has concentrated upon mapping the Moon. Cosmology is essentially mathematical, and I am no mathematician. This being so, it is natural to ask why I am writing a Foreword to this book.

The answer is quite straightforward. Despite his scientific qualifications, Hilton would be the first to agree that many of his views are completely unconventional. He does not believe in the "Big Bang" theory, according to which the entire universe— space, time, matter, in fact everything—was created at one definite moment in time, 13.7 thousand million years ago. In this he is not alone, and indeed the term "Big Bang" was first used scornfully by Sir Fred Hoyle, who was equally sceptical about it. But Hilton goes even further, and rejects the concept of an expanding universe, upon which all current cosmological theories are based. To him, instead of being immensely remote and immensely powerful, quasars are minor features expelled from relatively close galaxies; all our distance measures beyond the Galaxy are rejected for uncertainty. We must throw away most of our cherished thoughts and start again.

The instinctive reaction of many readers will be to give a sad smile, close the book and discard it. This, I submit, is precisely what should not be done. Hilton's theories, wildly unconventional though they may be, are backed up by what he regards as convincing evidence, and before rejecting them the reader must surely examine them very carefully indeed. It is easy to see that a tremendous amount of research has gone into the book.

Also, it is important to avoid unscientific prejudice. There is no doubt that Halton Arp, a leading observer but another Big Bang sceptic, was refused the use of large American telescopes because he was producing results that cut across conventional "sacred cows", and this is not an isolated case.

Is Hilton Ratcliffe right, or is he completely wrong? Probably 97% to 98% of modern cosmologists will say that he is wrong. I will reply differently, and use a well-worn catch-phrase: "You may well think so—I cannot possibly comment." But I cannot resist referring to a General Assembly of the International Astronomical Union in the early 1960s, which I attended as a member of the Lunar Commission. Quasars had just been discovered, and nobody knew exactly what they were. On one occasion, I "looked in" during a discussion about them, and asked an innocent question: "Would it be worth looking around major galaxies, such as Centaurus A, to see if there is a concentration of objects with quasar-like spectra?" I forget what was said, if anything, but it seems just possible that the question was relevant.

In any case, read *The Static Universe* before making up your mind.

Patrick Moore
"Farthings"
Selsey, West Sussex
August, 2009

Preliminary Notes

This is the most enticing challenge I can ever remember having on the table before me. The great danger in compiling an account of this particular subject—and it is a danger, I must confess, to which I no doubt succumbed—is that there is such an overwhelming abundance of evidence on every side favouring my thesis that it is almost impossible to resist the temptation to overstate my case. If I have once again produced a bloated tome, please pardon me. Perhaps I need professional help.

I should emphasise, however, that The Static Universe is not the progeny of my first book, The Virtue of Heresy, although they are in succession temporally, and share many basic tenets with one another. This book is slightly more formal and hopefully better behaved. Nevertheless, I do urge you, for reasons I coyly conceal, to read my earlier work as well. It makes some important points, and may just be the world's best introductory text to non-aligned physical science (or so say I).

Quite unintentionally, the current work has fulfilled an earlier promise, one which I would hope might have far reaching consequences for science and for the world: It has turned out to be the backbone of a worldview constructed from true observational cosmology, set in real space, and quite independent of models. The concept is a recurring dream, and could well inspire me to tackle the project more formally once this present effort is safely on your bookshelf.

Before we go any further, let me clarify the choice of words in the title. The word "static" does not imply that the Universe is standing still. It is a standard term with a specific meaning in cosmology: It means "non-expanding." The global pool of terminology relating to cosmology and associated fields of investigation is obese. I certainly don't wish to be cavalier in adding my 5-cents-worth to the list, but I quite unexpectedly distilled a principle from the analysis of data for this work that has to be named. It pertains to the sociological side of practicing physical science, and has been around for millennia as far as I can tell. It has had a profound effect on the development of the current Standard Model of Cosmology. I have named it Ideological Momentum. Here are these two key terms as they appear in the glossary at the back of this book:

> Static: In cosmology, an adjective qualifying the cosmos such that it does not organically expand; a static Universe is none of spreading out, becoming less dense, or growing larger.

> Ideological Momentum: The impetus of collective opinion; the tendency for supportive results to emerge and grow artificially from prior consensus or authority; also called "the snowball effect"; a synthetic trend in which we impute meaning in things just because we want meaning to be there for whatever deeply held reason, and then take that meaning forward even when it has been objectively falsified.

In summary, I am constrained in my writing by the following etiquette:

Firstly, Political Correctness. The term is insidious; it takes the word correct and removes the beauty of truth from it. It is not my intention to insult or patronise anyone, so if I do so, please pardon me. It wasn't premeditated.

Secondly, the conventions applicable to footnotes are the same here as they are in *The Virtue of Heresy*. Footnotes will cover some areas in more detail for those with a scientific bent, or allow me to add an aside, but they are optional reading. Their omission by the reader will not materially affect the flow of the story.

Thirdly, numbering follows the American convention—a billion is a thousand times a million, etc., commas delimit thousands, and integers are separated from decimals by a point. Quantities expressed scientifically use metric (decimal) units of measure. The *Système International* (SI) conventions are followed where applicable.[*]

Very large and extremely small numbers are often represented exponentially, for example one billion can be shown as 10^9 and one divided by a thousand as 10^{-3}. A convenient way of writing the reciprocal of n is n^{-1}. Apart from that, I hope I have managed to avoid the use of any rigorous mathematical expressions!

Fourthly, abbreviations and symbols abound! Be prepared for a deluge of acronyms, the *lingua franca* of astronomy. Classical conventions—the Greek alphabet, used extensively in science, and Latin (Roman) prefixes denoting scale—are listed in an addendum. For example, Λ (lower case, λ) is *lambda*, usually representing the cosmological constant, and GLY stands for giga light years, giga denoting a billion.

Fifthly, some of my personal quirks: The use of capital letters to denote proper nouns. It's Sun, Moon, Solar System, Big Bang, and Black Holes. The known universe—the extent of our large-scale environment that we are sentiently aware of and can measure—is spelt with a lower case "u." The hypothetical entity that represents the sum total of absolutely everything, and is probably infinite in every way, is called the Universe. I note with interest that in this book, where I am trying to deal with measurements, it is nearly always "u." In The Virtue of Heresy, on the other hand, it was mostly "U." It's a barometer of my shift in approach.

The set of nuances emerging from the measurement of motion in different frames of reference, irrespective of geometry and any particular co-ordinate system, is named relativity. The specific method authored by Albert Einstein, including the philosophy that resulted from it, is spelt Relativity.

Finally, no nonsense! I am not here to encourage any lascivious passion for dark, magical things that hiss from the shadows of our imaginations. Reality does not exist because we agree upon the idea; it exists, and therefore rational folk can interact sensibly within the constraints of the independently real world we fortuitously find ourselves in. Nature will provide the answers in her own good time.

---

[*] *Système International d'Unités*: A system of units adopted by international agreement for scientific and technical measurement. There are seven basic units – the metre, kilogram, second, ampere, kelvin, candela, and mole – and multiplying or dividing these seven obtains all other units.

So much of modern scientific thinking is open to a broad spectrum of interpretation. Whatever the underlying truth about the cosmos may be, it is always the effect of that truth that we see in the world around us. If this really were a Big Bang Universe, then all observations, without exception, would be evidence of that. Conversely, if the Universe is indeed static as I suggest, then no observation should show otherwise. It's all a question of finding the correct interpretation of the evidence, and there's no room here for fence sitting.

Often I've had to make a call, and I've done so in the full knowledge that sometimes the direction I've chosen will cause outrage in the minds of many of my generation's finest thinkers.

But do that I must, and I'm ready to take the heat.

Hilton Ratcliffe
Durban, South Africa, 2010

Chapter 1

Far Things

The red herring

"A man cannot strongly enough ask of Heaven: if it wants to let him discover something, may it be something that makes a bang. It will resound into eternity."

(Georg Christoph Lichtenberg 1742-1799)

"To hear scientists talk today, you would think the first moment in human history in which nonsensical views are not widely held is now."

(Sir Fred Hoyle).

♦ ♦ ♦ ♦ ♦ ❖ ♦ ♦ ♦ ♦ ♦ ♦

The Universe described in the Standard Model of Cosmology, for which we shall indulge the colloquialism Big Bang Theory, does something rather strange. It *expands*. Every large object in the modern projection of the cosmos moves uniformly and systematically away from each and every point of observation, irrespective of where the observer might find itself. From where I am standing, however, this doesn't appear to be the case. Is there any reason other than weight of authority to hold in good faith that the entire Universe is systematically expanding? I believe not, and you have before you my account of events and people that brought us this far down an enticing but wholly misleading garden path.

A great contemporary alderman of observational astronomy (I shall with regret abide by his wish for anonymity, for I owe him so much) suggested this book to me of a cosy winter's evening in his ancient, book-lined study. He was quite definite; I was to focus specifically on expansion, and furthermore, to pivot it upon the issue of redshift.[*] He's right of course, and I shall attempt to do that, but as you will soon see, the interdependence of celestial phenomena is so beguiling that I was before long compelled to loosen those constraints just a tad. By the time we finally arrive at chapter eight, you might well wearily ask what on Earth something as arcane as gravitational lensing has to do with universal expansion; I should hope by then that the answer may be given, and your patience rewarded.

The principal issue in my opinion, the one upon which Big Bang Theory lives or dies, is universal expansion. It is nothing less than an axiom in space science. It is taught as fact in universities, and assumed, nay, *expected* by referees and editors vetting contributions to the journals of science. Universal expansion is believed implicitly by the majority of investigators in the field. There is no public debate on the matter. Why should this be so? What overpowering evidence convinced so many talented people that every analysis of a celestial object just *has* to be seen against the background of a receding cosmos? I really needed to find the answers to these questions, because my own observations, and a great many others besides, were clearly telling me otherwise. Had I missed something? This book is an account of my investigation, and whether or not you concur with my conclusions, I'm sure you'll agree with me that answers I found are truly astonishing.

What is cosmology? It is a preferred theoretical treatment of the large-scale biographical properties of the universe. The noun cosmology derives directly from cosmos, which is the Universe considered as a system. Something described as cosmological is implied to be both non-chaotic and non-local, that is, both orderly and vast. The current paradigm was born with the publication of the General Theory of Relativity (GRT) in 1915, which in turn was a consequence of the earlier Special Theory of Relativity (SRT). In

---

[*] Cosmological redshift is a deflection of dark lines in the spectrum of starlight, measured against the benchmark of sunlight. It indicates dropping energy (increasing wavelength), usually taken to mean that the light source is receding from us.

a remarkable set of visionary publications, Albert Einstein commenced in 1905 to completely revise our understanding of light and space. In so doing, he impressed the ideal of non-Euclidean geometry upon young astronomers like me who would nearly a century later trail in his gleaming wake. It was nothing short of a revolution.

Cosmology was thus removed from gnarled old hands twiddling knobs on telescopes, and henceforward became the exclusive domain of those select few who could understand and solve complex differential equations. What they conjured up in that private theatre of the mind would henceforward determine how we were to see things. Collections of vectors called tensors, held in fragile relief in the mind's eye, replaced the central role of optical images in determining cosmological reality. The fundamental tenets of large-scale astrophysics were transformed from what is seen to what someone thought up, and the ancient Ptolemaic principle of pure reason dominating what we experience was summoned from its musty tomb to once again become supremely respectable.

I shall be referring to strange acronyms, repeated abbreviations, and some unusual terms so frequently in the pages before us that it would be as well to familiarise ourselves at the outset. This is a brief review of what we know of the Standard Model of Cosmology. From the womb of General Relativity Theory (GRT) came three models, engineered from the equations by youthfully brilliant Russian mathematician Alexander Friedmann. To Einstein's initial bitter chagrin (for he envisaged a static universe after all), the Friedmann solutions strongly suggested that the Relativistic cosmos was indeed globally dynamic, and should in fact be expanding in all directions. A Belgian cleric named Georges Lemaître reasoned that in some time past, expansion must have commenced from something inconceivably dense, and charismatic Russian émigré George Gamow gave the whole concept formal plausibility—and a deal of showbiz pizzazz—by adding stupendous heat to the equation. In a nutshell, this is the expanding universe picture currently known as the Lambda-Cold Dark Matter Model (abbreviated Λ-CDMM or LCDMM).

Let's clarify a few things here. *Lambda* (in Greek notation, Λ or λ) refers to a repulsive gravitational effect, a negative force counteracting the collapse of the Universe so vigorously that it is said to be blowing everything apart. It's commonly called Dark

Energy. Opposing this Dark Energy is an attractive impetus ema-
nating from Dark Matter, described as "cold" in the model because it does not radiate. Like Dark Energy, Dark Matter is completely invisible. Because it manifests as halos around large, visible objects, Dark Matter must also be absolutely transparent. Astoundingly, observed or measurable phenomena that can be addressed by physics account for less than 10% of the LCDM cosmos. More than 90% of cosmological agents in Big Bang Theory
are supernatural, unprecedented, and invisible. Electrophysicist
Tom Wilson puts our astonishment better than I can:

> In an interesting philosophical aside, if 96% of the Universe
> is unobservable dark matter and dark energy, then why
> bother looking at the real thing any more? Perhaps this is
> the unfortunate logical dead-end to a Λ–CDM model. [*]

No one has conquered this domain, so let's try to be bold. It's
given that uncertainty increases with remoteness on all axes. In
my view, those claiming definite answers to distant problems are
in all likelihood misled by imperfect methodology. We should
check the conclusions already reached in space science by inter-
rogating the techniques used in their formulation; reject those
that have been falsely proclaimed (no matter what the cost to our
favourite theory); apply our minds to the devilishly difficult task
of developing other empirically verified means of measuring
things that are far, far away; and work with reality-tested physics
to decipher the coded messages that have travelled across the
aeons of space to puzzle us here on Earth. In this way, we may in
the pages before us, mitigate the unseemly haste and synthetic
science evident in adopting and adapting observations to suit a
preferred model, and also—hopefully!—reduce the rampant ignoring of evidence that runs contrary to those prior preferences.
The foregoing paragraph introduces the mission statement of this
book; the next puts it in perspective.

> The genuine rationalist does not think that he or anyone
> else is in possession of the truth; nor does he think that
> mere criticism as such helps us achieve new ideas. But he
> does think that, in the sphere of ideas, only critical discus-
> sion can help us sort the wheat from the chaff. He is well
> aware that acceptance or rejection of an idea is never a
> purely rational matter; but he thinks that only critical dis-

---

[*] Tom Wilson, "Dark Matter Recreations Part Two, September 30, 2009"
(www.thunderbolts.info /tpod/2009/)

cussion can give us the maturity to see an idea from more
and more sides and to make a correct judgment of it.

These are the words of acclaimed philosopher of science,
Karl Popper. The man who gave us the mantra that all scientific
theory should be capable of falsification, left us in addition this
sage advice on the nature of intransigence:

> Whenever a theory appears to you as the only possible one,
> take this as a sign that you have neither understood the
> theory nor the problem which it was intended to solve.

Upon sober analysis, it would appear that the Standard
Model of Cosmology is not nearly as bulletproof as the bulletins
would lead us to believe.

We duly acknowledge the uncertainty that follows great remoteness in time or space, and go forth under no illusion that the
path ahead is fraught with difficulty for all of us. Anyone claiming to *know* what happens at the end of the cosmological rainbow
should be treated with a great deal of caution, a caution that
should follow us evermore when we are greeted with absolutes. I
propose that we try an approach that takes note of uncertainty:
We give observation and experience precedence over theory. If
theory suggests A, and observation indicates B, we should lean
towards B. Theory will *emerge* from experience, not precede it. We
may usefully transmute from chemistry to physics a fundamental
principle: We analyse a reaction without in the least bit taking
into consideration which particular brand of space surrounds it.
Now that, I fear, is going to drive the esteemed fathers of cosmology to apoplexy, but I know also that the merits of such an approach will be clearly evident, if only there is willingness to try it.

Our method might profitably emulate other practitioners of
physical science, for example, neurosurgeons or structural engineers. Before they propose a procedure, they take enormous
trouble to ensure that it is operationally viable in the real world
of brain tumours and suspension bridges. Let us analyse the *behaviour* of organisms under review without colouring the picture
in philosophical nuance. Reduce the problem to operational effects and treat it accordingly, as if we were cosmic neuro-
engineers creating theoretical artefacts only if we were confident
that lives could depend upon them. Our own lives even. That
would very quickly eliminate wild theories and abstract conjec-

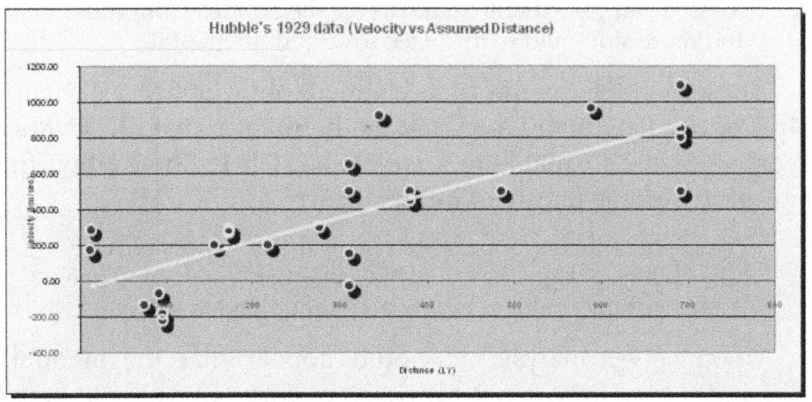

Figure 1: The plot of redshift against distance for Hubble's original galaxies. The preconceived assumption of a Doppler shift in this data set was later found to be baseless, and theory had to be revised to exclude "local" galaxies from expansion.

ture about what type of curvature to assume.[*] Of course, observer effects—biases in the quality and quantity of what is seen, usually aligned with the standard of resolution—are real enough, and must be taken into account, but with the caveat that, unlike prejudice, they apply as much to observers who would seek to support a preferred theoretical model as would deny it.

Please indulge me for a moment so that I can explain to you how I look at things. So that we are under no illusion right from the start, let me confess that my approach seeks to reverse the prevailing order of priority. I am an astrophysicist specialising in celestial mechanics and chemistry. That means that my approach to science is physical. My research is centred on observations obtained by astronomers, those utterly magnificent images in every useable wavelength of light acquired through the apertures of fantastic telescopes. In other words, my analysis begins with what is seen.

To the layman, especially those engaged in astronomy for the love of it, this may seem to be standard practice. It is not. Cosmology as it is currently performed does not proceed from observation; it commences from theory, expressed in exotic, convoluted mathematical dialects, and based upon principles believed to be true irrespective of observation. The most crucial of these to our

---

[*] The issue of space curvature is dealt with fully in chapter eight.

present discussion is the assumption of various types of space, and the influence that might have on how we see things.

As if that were not bad enough, I also refrain from using mathematics for any purpose other than to calculate and express quantities. Although mathematical modelling is unarguably a powerful and innovative tool that has produced some marvellous, useable physical artefacts—the harnessing of nuclear energy and lasers are prime examples—it is unfortunately also capable of convincing us to camp in blind alleys. Theoretical structures eventually become so convoluted, and the means of expression so arcane, that they end up as self-fulfilling prophecies almost impossible to refute on their own terms. Debates invariably lose themselves in arguments about syntax and the meaning of terms, ultimately blurring their somewhat tenuous connection to physical reality, and protagonists are often seduced by nothing more than the sheer elegance of mathematical expression.

If I may, I should like to quote from one of the anchors of modern cosmology. In 2001, Stephen Hawking published a sequel to his best-selling book A Brief History of Time. It's called The Universe in a Nutshell, and in it he says (on page 31):

> ...a scientific theory is a mathematical model that describes and codifies the observations we make. A good theory will describe a large range of phenomena on the basis of a few simple postulates and will make definite predictions that can be tested. If the predictions agree with the observations, the theory survives that test, though it can never be proved to be correct. On the other hand, if the observations disagree with predictions, one has to discard or modify the theory.[*]

Why should contemporary cosmologists consider themselves exempt from this?

At no time in the long, blood-stained history of cosmology has there existed a dominant paradigm that has not by the investigation of a few resolute individuals been proven partially or completely wrong; and nowhere do we find a record that such a paradigm was not vigorously defended with contrivances, even in the face of incontrovertible evidence that would render it false. What's more, every successful cosmology laboured under the misapprehension that what it described was the entire Universe. Are we now to believe that the currently dominant Standard

---

[*] Stephen Hawking, *The Universe in a Nutshell* (London: Bantam Press, 2001).

Model of Cosmology is the final answer? That it is beyond fundamental criticism?

There is absolutely no need to be nervous. This whole process should be accomplished without hardship, although I surmise that many of my readers would not necessarily have had prior insight gained from formal, graduate education in astrophysics or related fields. I must emphasise that you are in no way excluded from an understanding of the concepts. In most cases, a good dose of common sense will be more helpful in grasping the principles involved than a degree major in relativistic physics. No one in science has privileged vision. When you look at an image of a star, or when I look at an image of the same star, or when Albert Einstein or Stephen Hawking looks at it, we all see the same thing. None of us, I believe, has divine privilege in this regard. Yet what we make of it may be so vastly different that we are left wondering whether we were in fact examining the same object.

If we have an everyday understanding of the world in our immediate neighbourhood, we can comprehend the stars above. They are made of the same few elements, in the same way, under the same set of rules, as the familiar creatures we share our gardens with. Despite what the bigwigs tell us, there is no reason to believe that a completely different mindset, or indeed, entirely different physics are required in astronomy.

Please don't think I'm being flippant or facetious. It's true. Michael Faraday, the 18th century self-taught experimentalist, whose apprenticeship in the trade of bookbinding at the age of 13 saw the end of his schooling, went on to become one of the most influential figures of all time in the fields of physics and chemistry.* Faraday used observation and experiment to gain an understanding of often invisible processes of nature, and in that way provided the basis for those who followed, most particularly the father of electromagnetism, James Clerk Maxwell. The Faraday Effect, relating the interaction of electricity and magnetism to rotation, led to the production of dynamos and electric motors, with profound significance for the emerging technological era. He saw more of the real Universe in a candle flame than anyone

---

* Michael Faraday gave a famous series of "Christmas lectures" in London in the mid-19th century, and those of 1861 were transcribed into the book *The Chemical History of a Candle* (New York: Crowell, 1957). He was honoured with 95 degrees and orders of merit, and by decree, inducted as a Fellow of the Royal Society. He had no formal education beyond primary school.

ever wrung out of poly-dimensional geometry. Michael Faraday let theory emerge from observation.

In this work, we propose that the Universe is *static*. In the terminology of cosmology, that does not imply that it is standing still. The term "static" refers to an absence of any global, all-encompassing motion, and is usually taken to mean non-expansion. Basically, this means that the Universe is not evolving, in other words, that it is not as an entire organism advancing from a primitive state. The cosmos is indeed composed of component systems which are magnificently dynamic, especially in their rotation, but all we see is phases of innumerable local cycles. There is no indication at all of an overall, ongoing dispersion. From observation alone, we have identified the notion of universal expansion as the abiding, central flaw in modern cosmology.

Each of us is rewarded with the same picture when we look up at the night sky: Points of light by the tens of thousands, each an individual rendition of a remote celestial object. These points on the cosmic map should be flying apart if standard theory holds, yet perceptibly, they are static, maintaining by and large a fixed spatial arrangement. Perhaps they are too close to be expanding, or too far for it to be noticeable; whatever, we all see the same thing when we look, even with the most powerful instruments on Earth. All we see is a static Universe.

So it seems that we owe ourselves a really searching question—if observation indicates a static Universe and theory dictates expansion, how do we reconcile the two? We might argue that expansion occurs only beyond what we have been looking at, and indeed, that is what current theory suggests; or, on the other hand, that what we are seeing is simply too remote for the divergence to be visibly appreciated. In the latter case, we need to be very careful of circularity. The Standard Model of Cosmology assures us that expansion exists despite what we see, and that we are misled *inter alia* by the fact that these things are very, very far away. But the remoteness of those objects is given by a property of expansion itself—cosmological redshift. The argument collapses by being self-referential. The Standard Model attempts to prove expansion by assuming expansion. Our aim with this work is to expose that as extremely poor science.

Mathematical cosmologists sometimes suggest that we do not see expansion because we project it mentally into Euclidean space, whereas in reality the image should be seen in Minkowski space, or Riemann space, or de Sitter space for us to get the true picture. This presents great difficulty for us. User-defined reality is an anathema to me, and I should hope for you too. What I propose, therefore, is that we ignore that particular parameter. Let's simply take the image before us for what it is, like an apple in our hand, and leave aside the notion that in another kind of space, in a more esoteric mode of expression, it is actually, say, a frog. The intrinsic "apple-ness" of the object being observed is completely independent of the fact that you and I may be describing it, and as for its alleged "frog-ness", all I can say is show me or I'm just not going to believe you. To those belonging to the faith of mathematics, my suggestion is nothing less than outrageous heresy. Why is that?

The root of the current crisis in science is that we physicists so easily and so completely believe that the mathematical component in our education endows us in some mysterious way with a deeper and more profound understanding of nature. I don't believe it does. It is my firm conviction that nothing in astrophysics except quantities need be expressed in mathematics exclusively, and surely, nothing belongs solely to unapproachable elite. In conversation with Virginia Trimble, eminent astrophysicist Rocky Kolb summarised the challenges of cosmology like this:

> Our goal must not be a cosmological model that just explains the observations, the ingredients of the cosmological model must be deeply rooted in fundamental physics. Dark matter, dark energy, modified gravity, mysterious new forces and particles, *etc.*, unless part of an overarching model of nature, should not be part of a cosmological model. We may propose new ideas, but they must wither unless nourished by fundamental physics.[*]

The Universe as we perceive it is a hierarchy of systems, percolating up from beneath micro-atoms to way beyond macro-galaxies, and it would appear that they are all somehow linked in a gargantuan strategy. Chemically at least, the rules governing reactions are the same near and far. Emerging compounds are always constrained by these rules, a sublime continuity that paints the entire known Universe from the same small palette of colours. This presents a clear opportunity to analysts like me, especially as I am endeavouring with this work to craft a useful

---

[*] Cited in Vicent Martinez and Virginia Trimble, "Cosmologists in the Dark" *arXiv*: astro-ph/0904.1126.

Figure 2: The M80 globular cluster of stars embedded in the Milky Way. Theory exempts local space from expansion, so it cannot be properly tested. Any optically visible evidence of expansion occurs beyond the detection horizon. (Hubble Heritage image courtesy of NASA/AURA/STScI).

précis of realistic cosmology. The only fair test of a scientific theory is against physical reality.

Cosmologists and unfortunately even astrophysicists are driven by preferred beliefs. They will take a stand on the merest whiff of evidence. That is simply not good enough, I'm afraid, and science is in trouble because of it. The expansion, they tell us, is beautifully symmetrical. It seems immune to the chaotic deflections and overwhelming tidal effects that would characterise an explosion of real things.

The Universe may not be as it appears to be. We have been deceived before by optical illusions, and experience confirms the need to expect the unexpected. It is hidden mystery rather than the comfort of plain sight that defines the allure of cosmology. "Science is the culture of doubt," said Richard Feynman. The great man was right; we seek clarity in a whirlpool of uncertainty. He also said that if we could not explain a theory to our grandmother in a way that she could easily understand, then we did not understand it ourselves. There's more than a grain of truth there; unfortunately, unless our grandmother is particularly well-educated in science, it clearly rules out differential geometry and tensor calculus, doesn't it?

The Big Bang model, in all its components and incarnations, is an extraordinary hypothesis, and the burden of proof lies squarely on the shoulders of those who would promote it. With that in mind, we should not forget that the model is not in any way an extrapolation of what has been observed. On the contrary, it was compiled from isolated intellect prior to any observational support being found. The "expansion" is not a mechanical process; it is a way that spatial dimensions roll out according to one particular solution to a specific set of equations. And really, I have to say this; the precise, uniform way it is supposed to expand stretches one's credibility to the limit.

I shall now give you a thumbnail précis of the expansion model that we will be dissecting and analysing in the pages ahead. Take a deep breath. Here is how it works: The Universe in standard, consensus theory is contained by space that is constantly increasing in volume. This creates the appearance of expansion, with large objects moving away from each other. The rate of expansion is everywhere the same, so that any two neighbouring objects always depart each other at the same speed. From any point of observation, the expansion is radially uniform and progressive. We add each local rate of departure to the next, so that recessional velocity becomes smoothly greater as distance from the observer increases.

This effect has become known as the Hubble law. It is said to apply only at some arbitrary "large scale" (the Local Group of galaxies, for example, was found to be mysteriously excluded from expansion), and it is calibrated by a scale factor known as the Hubble constant or Hubble parameter (H_0). It is a simple parameter, expressing speed per unit of distance, thus:

> H_0 = n kilometres per second per Megaparsec (where n is the value assigned to the parameter, "kilometres per second" is the unit of speed, and "Megaparsec" a unit of distance equal to approximately 3.26 million light years).

As an example, let's take a currently popular value for H_0, about 70 km sec per Mpc (the consensus value varies between 50 and 100). Applied to the expanding universe, it means a galaxy 1 Mpc away from us should recede at 70 kilometres per second, another galaxy 2 Mpc away, at 140 km/sec, and a third galaxy a further megaparsec more remote, at 210 km/sec, and so on, forever. It should be mentioned before we conclude our introductory sketch that according to the model, "gravitationally bound" systems do not contain expanding space. Therefore, space between the stars here in the Milky Way is static. It only starts expanding some distance away. No one appears to know quite where that far away place might be.

That's enough detail for now. We have a general idea what theory predicts for the cosmological environment. Now, in the cold light of day, how do our measurements stack up against the model? We are surrounded by the Universe. It behaves according to the laws of nature. My job as a physicist involves determining what those universal principles are, thereby developing an understanding of the world as I see it. My first clue is what I observe and experience. Let's try to put the theory of expansion into a real-world context.

As I sit here writing this book, the Universe cradling me is represented and contained by the four walls of my study. Should you make so bold as to suggest that the room is expanding, that the walls are steadily and progressively moving away from me, I shouldn't reject the idea out of hand. I should, however, be circumspect. It's an extraordinary idea, because as far as I can see the walls are static, and it is unlikely in the extreme that they would just spontaneously start flying away.

What is required is healthy scepticism. I would expect you to qualify your theory with some hard measurements. If you cannot do that small thing, don't be surprised if I continue to conclude that the room, and the house, and the garden outside, are not appreciably spreading out. You will have failed to convince me that I am deceived by my eyes.

The evidence cited in support of the extraordinary notion that the visually static Universe is holistically expanding is extremely weak, to the extent that it might be called non-existent. In fact, as we shall see, the original dataset from which the supposition of galaxy recession was drawn was within five years of its announcement rejected as a mistake, and the pioneer who discovered it distanced himself completely from the idea. It's a fascinating story woven around the power of belief and prejudice in human nature.

It's a mindset problem, I would say. What's the answer? How do scientists deal with anomalies? Their response is so consistent and so predictable that Thomas Kuhn made it a law. Kuhn's classic, The Structure of Scientific Revolutions, was described in The Times Literary Supplement as "…one of the hundred most influential books since the Second World War."

He pulls no punches:

> Part of the answer, as obvious as it is important, can be discovered by noting first what scientists never do when confronted by even severe and prolonged anomalies. Though they may begin to lose faith and then to consider alternatives, they do not renounce the paradigm that has led them to crisis.[*]

True to form, they still, after all the evidence to the contrary, cling incorrigibly to the notion of an expanding Universe.

The spiritual father of this currently dominant paradigm, the good abbot Georges Lemaître, told us in 1924 that fiery creation issued forth from what he euphemistically described as a "primordial atom", which somehow exploded and drove the galaxies apart. Although his charmingly naïve version of events has been superseded in the interim by magnificent theories and daring science, one unflinching aspect of the hypothesis remains steadfast to this day: The Universe, they tell us knowingly, is beyond any shadow of doubt, *expanding*.

It must be said, once again and not for the last time, that we do not see an expanding Universe. Overall, the cosmos we observe is apparently static, and, it would seem, in an elegant hierarchy of gravitational equilibriums. For as far as we can make out, there is all around us balance between the poles of opposing forces, and there are no data to indicate that it may somehow be different over the horizon of our ability to take measurements. Of course, this balance fluctuates, but after every violent breaking of the symmetry, we see things settle down again to the calmness and serenity of polarised balance. Supernovae blast material wildly into space, but just look at the symmetry that eventually

---

[*] Thomas S. Kuhn, *The Structure of Scientific Revolutions* (Chicago: University of Chicago Press, 3rd Edition, 1996).

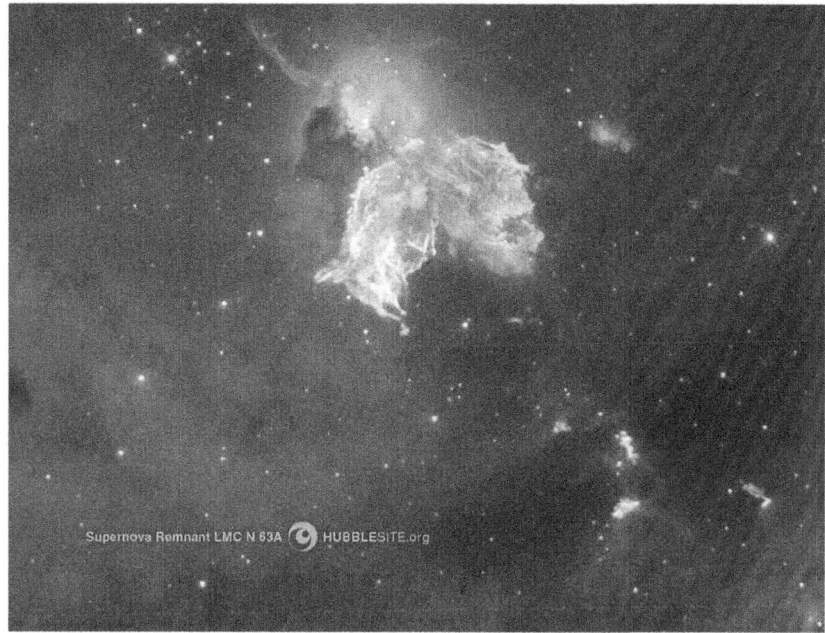

Figure 3: Chaotic aftermath of a catastrophic stellar explosion called a supernova. Why doesn't the Big Bang Universe look like this? (Image courtesy of NASA, ESA, and STScI).

results. Our Solar System, supremely elegant, emerged from the lashing tides of a supernova's nebula.

The Universe in which we live and breathe is not chaotic; just look up at the sky on a clear night, and as your eyes grow accustomed to the dark you will see for yourself the tiniest part of this magic at work. The stars are not simply peppered across the heavens, lying where they fell with no consideration for their neighbours. The band of light dissecting the dome above is living proof that those stars are organised. It is not proletarian anarchy. No, it is an ordered Universe, even though it might be suffering a prolonged bout of hiccoughs. There is without doubt a scheme, precisely followed, that allows our stuttering cosmos to take knocks and to enjoy recovery. The key, I surmise, lies in understanding the role of many small explosions in the rollout of galaxies, and the function of spin in sustaining equilibrium.

The finest theory in the world remains singularly unimpressive, in my book, if it is not clearly and unambiguously supported by measured quantities from our observed environment. The technical aspects of expansion theory are profound, and the ways that measurements are made are necessarily complex and daunting, but please bear with me. I shall try to make the process as painless as possible.

We should first examine what Big Bang expansion theory requires of a universe constrained by such hypothesis, and then see if our experience and observation might support its existence. The requirements include:

- A completely smooth, symmetrical distribution of matter and material structure (that is, homogeneous and isotropic—the Cosmological Principle).
- Large objects moving progressively away from one another; yet simultaneously, the existence of *no* large objects as required by the Cosmological Principle.
- A mechanism to drive expansion, in this case the creation of spacetime between gravitationally bound systems.
- Finite limits to space and time.
- Redshift increasing uniformly with distance.
- Evolution of structure correlated with redshift.
- Evolution of all chemical elements and forces from particles.
- A universal, ongoing, linear process of bottom-up, large-scale structure formation.
- A solution to the horizon problem (Inflation Theory).
- A uniform, surrounding, radio wave picture of the very early universe.
- Complete consistency with Special and General Relativity.

Not a single one of these requirements is met. I repeat: *Not one.* I shall attempt in the chapters that follow to explain in plain language my justification for such a damning statement. Although the expansion hypothesis rests on only two observational tenets—being principally, cosmological redshift and by intricate inference, the microwave background—there is more to my challenge than that.

As these "observations" were made, it became clear that they exacted a heavy price on physics. The overhead was such that at each step along the way, *ad hoc* theories had to be invented to make it work. First was the replacement in cosmological redshift theory of the Doppler Effect[*]. It was untenable, but it seems to me that the replacement theory—the creation *ex nihilo* of spacetime, supposedly with zero-point quantum energy, between (and only between, not inside) astrophysical structures—is vastly more so. It flies in the face of the laws of thermodynamics and the conservation of energy, tenets so fundamental to physics that their breach is simply inconceivable to sober scientists.

Then, in the 1960s, came the Cosmic Microwave Background Radiation, a phenomenon that just couldn't be explained without Inflation Theory. Here the fathers of cosmological reasoning excelled in the absurd. I shouldn't think any other commonly held theory in the history of science has surpassed the flagrant disregard for physical reality and rational thought expressed in Inflation Theory. To suggest that expansion could fly to the piper's tune, at superluminal speeds, and vary its acceleration without known or conceivable cause, precisely as required by the model, is utterly ridiculous.

After that, the rest of the Lambda-Cold Dark Matter Model, positing as it does that what is seen can only be explained if more than 90% of the energy manifested in space is supernatural, undetectable, inert in every sense excepting gravitational attraction or repulsion, and also completely transparent, seems tame by comparison. Alas, we have thus become desensitised to the cavalier mischief of Hollywood-style science.

However, if redshift means distance, and therefore old light coming from a "young" Universe—that is, one where objects are tiny, faint, immature, densely packed, and at very high temperatures—then we know what to expect in observations of high redshift objects. They should appear smaller, duller, younger, closer together, and hotter than comparatively low redshift objects. In the chapters ahead we will list observations with which we can test our expectations. I am sure you will find them fascinating. In order to fulfil the promise of this book, I am going to have to present argument that:

- Expansion as proposed by Big Bang Theory invokes a complete void in physical theory to suggest creation of space itself (essentially, the creation of energy);

[*] The Doppler Effect is seen in the change in wavelength of signals coming from a moving object. Redshift is a lengthening of the wave from a receding light source.

- The Universe appears to be infinite. There is nothing indicating that it is finite. Olbers' paradox carries no weight in countering an infinite, cyclic, static Universe;
- The *ad hoc* imposition of inflation defies established physics;
- The Hubble law is a fallacy; in any case, a static Universe can present the Hubble redshift, in a variety of ways;
- The Cosmic Microwave Background is simply a diffuse image of local astrophysical structure at the equilibrium temperature of starlight;
- A static Universe can exist without collapsing, and a non-expanding Universe does not contradict the observed abundances of elements; on the contrary, it succeeds where an expanding, evolutionary Universe model has so far failed;
- Linear evolution of structure with time cannot reasonably be accommodated within the limits of the Big Bang era, and is in any case neither correlated with redshift nor indicated by deep field surveys; observed exceptions show that higher redshift objects are not necessarily less mature, less bright, and closer together, or crucially, further away, than those with lower redshift values.

Time out. Let's deal with the word "creation" before we proceed. It crops up all over the place. This is because the notion of creation is hard-wired into Big Bang Theory, both in implied origin in time of the complete Universe and in the increasing-space hypothesis that drives expansion. It vexes many of us engaged in the pursuit of natural philosophy, and at times some even feel the need to become almost apologetic. Alan Sandage in his brilliant review Observational Tests of World Models takes a stab:

> Creation is a flammable word that triggers responses often not intended by writers who use it ... Nevertheless, the subject is as close as science can come to the questions of origins—hence its enormous appeal.[*]

What I would like to say on the matter is simply this: Despite carefully contrived tap dances and cunning diversions, the L-CDM model clearly does postulate a finite beginning in time, and thus creation in the sense that something appears where there

---

[*] Alan Sandage, "Observational Tests of World Models" *Annu. Rev. Astron. Astrophys.* **26** (1988), pp. 561-630.

was nothing. The same goes for the creation of spacetime (with zero point energy) to drive expansion. Both are quite impossible. Creation, whether of the entire Universe or just a part of it, cannot proceed from nothing at all. Every created effect must necessarily and without exception be given by a creating cause. Since this is logically true for all cases, the infinite Universe is proven. I leave it up to you to put your own preferred label to that prior and eternal progenitor.

Finally, we cannot ignore the role of education. At graduate level, we start to apply ourselves to problems of science using well-worn, honed procedures we have been taught to use. Speaking as an astrophysicist, it seems most scientists accept the Standard Model, not because they have thoroughly tested it, but because *that is the way they have been trained to do science.*

We ought not to apologise for this—there is nothing inherently wrong with doing things we've been taught to do. But in the interests of proper science, we should never forget to ask questions and expect answers. A flawless educational curriculum has yet to be invented. Let's take just a moment to talk about this. It's important.

We graduate with a set of tools in our kit. We eagerly anticipate using them as we fearlessly dissect the Universe. One of these is a compendium of *standard models*, each a description of a particular substratum, and each based upon the prior work of leaders in the field. They are written into the book of science by consensus (although consensus by *whom* is a chilling question).

Thus we have as examples, the Standard Solar Model, the Standard Model of Particle Physics, and what we are talking about here, the Standard Model of Cosmology, properly known in its current configuration as the Lambda-Cold Dark Matter Model. As you no doubt noticed, I am cheeky enough to echo Fred Hoyle's amusingly accurate title, Big Bang Theory.

Let's not be naïve. It is imperative that we work off the efforts of scientists that preceded us or we will simply stall in the starting gate, sucked into a vortex of eternally reinventing the wheel. However, we must in some way be convinced that the prior results and speculations are sound enough to anchor the progress of our entire careers, for one bad choice would be disastrous. Therein lies the rub; we are invariably persuaded by nothing more than the authority and reputation of our legendary predecessors. The rock star status of Albert Einstein in the 1920s and

Stephen Hawking today was driven by fans that, with respect, hadn't a clue what those gentlemen were actually on about.

My experience and interaction with many fine, conscientious investigators engaged in space science over the last three decades has shown that Big Bang Theory is almost always accepted without question by undergraduates, or if they do have doubts, these are suppressed in the very sensible interest of passing an exam and getting a degree.

That's how the die is cast. In graduate school, the conditioning is further reinforced by the practical need to move forward, with our studies and with our careers. By the time we have achieved a doctorate and gone into teaching or research, it's done and dusted. The standard models are entrenched in the way we do science, and while it is surely not the principle that is at fault, it is nevertheless the *way* that models are canonised that leaves gaping wounds in the pursuit of physical science.

The joke is that it is only when we retire, or are dismissed, or disenfranchised, or in some other way become officially useless to the world that we can finally declare openly what all those years have shown us: We were little more than a pack of dogs, barking up the wrong tree. Our track record proves it—all the billions spent on searching for elusive shadows over the last half century, more often than not with cocked ears and wagging tails an attempt to reinforce one of our favourite models, have not produced anything that could be called a fundamentally new discovery in physics. That really ought to be telling us something.

My aim with this book is to reset the hierarchy of celestial science. The correct order of things, in my opinion, is astronomy—physics—theory. Our first clue, the first rung on this ladder of exploration, is the image. The picture acquired by astronomers is where we start. All knowledge of the cosmos should emerge from observational astronomy. Unfortunately, this places modern cosmology, created as it is in the insulated mental universe of imaginative thinkers, right at the back of the queue: In my scheme, it's astronomers first, astrophysicists second, cosmologists distant third. And mathematicians nowhere in sight! You can just imagine how appalled cosmologists are at the prospect of losing their prominence. They do not suffer me gladly.

I'm under no illusion how difficult my given task will prove to be. To convince really clever people that they are victims of their own cleverness, and that I am not, may prove to be a barrier

too resilient for our common sense after all. Don Scott has suggested that we are in any case not up against baffling science in
cosmology but simply confronted with human nature. Martín
López-Corredoira says it thus:

> Certain results of observational cosmology cast critical
> doubt on the foundations of standard cosmology but leave
> most cosmologists untroubled. Alternative cosmological
> models that differ from the Big Bang have been published
> and defended by heterodox scientists; however, most cos-
> mologists do not heed these. This may be because standard
> theory is correct and all other ideas and criticisms are in-
> correct, but it is also to a great extent due to sociological
> phenomena such as the 'snowball effect' or 'groupthink'.
> We might wonder whether cosmology, the study of the
> Universe as a whole, is a science like other branches of
> physics or just a dominant ideology.[*]

You may well point out then that cosmology, being at best
conjecture and subjective opinion, is neither here nor there, of no
great consequence to physical scientists, and you'd be right. In
1997, acclaimed French astronomer Jean-Claude Pecker expressed
scepticism many of us now share, whether we are free to declare
it not:

> And we would pretend to understand everything about
> cosmology, which concerns the whole Universe? We are
> not even ready to start to do that. All that we can do is to
> enter in the field of speculations. So far as I am concerned, I
> would not comment myself on any cosmological theory, on
> the so-called 'standard theory' less (than) on many others.
> Actually, I would like to leave the door wide open.[†]

Dealing with the subjective residue of scientific contemplation is hazardous. One has to be painfully aware that beliefs, of
whatever nature, are invariably the consequence of some sort of
prior, non-negotiable authority; someone, somewhere in the past,
had privileged vision and our faith in that vision is largely determined by our awe of his or her reputation. A holier-than-thou
elitism is invariably the consequence of belonging to the dominant tribe. We should address it only by means of a single question: The key is not *whether* we believe, or *what* we believe, nor

---

[*] Martín López-Corredoira, "The Sociology of Modern Cosmology" *arxiv*: astro-
ph/0812.0537.
[†] Jean-Claude Pecker, *J. Astrophys. Astr.* **18**, 481, cited in Martín López-
Corredoira, "The Sociology of Modern Cosmology" *arxiv*: astro-ph/0812.0537.

even *how* we believe, but simply *why* we believe what we do. By what process of reason, if any at all, do we confirm our faith? The answer to this question is surprising to even the most diligently introspective scientist, if only he would dare to ask it!

> It is not likely that we primates gazing through bits of glass for a century or two will dissemble the architecture and history of infinity. But if we don't try we won't get anywhere. Therefore we professionals do the best we can to fit the odd clues we have into some kind of plausible story. That is how science works, and that is the spirit in which our cosmological speculations should be treated. Don't be impressed by our complex machines or our arcane mathematics. They have been used to build plausible cosmic stories before—which we had to discard afterwards in the face of improving evidence. The likelihood must be that such revisions will have to occur again and again and again.[*]

In his book Cosmos, the late Carl Sagan laid out two rules of science:

> First: There are no sacred truths; all assumptions must be critically examined; arguments from authority are worthless.

> Second: Whatever is inconsistent with facts must be discarded or revised. We must understand cosmos as it is and not confuse how it is with how we wish it to be. The obvious is sometimes false; the unexpected is sometimes true.

Let's take a look at the world and see if we can spot any expansion, anywhere. We should find this interesting, but first, if you will, let us take time out to try to understand how Edwin Hubble visualised the cosmos.

> ***Discussion:*** *The Standard Model fails every reasonable test of science. It flies in the face of the laws of thermodynamics and the conservation of energy, tenets so fundamental to physics that their breach is simply inconceivable to sober scientists. The assumptions of the Standard Model are arbitrary, tuneable, in conflict with observation, and supported only by great uncertainty.*

---

[*] M. J. Disney, "The Case Against Cosmology" *arxiv*: astro-ph/0009020.

We have no good reason to believe what we do. No coherent philosophy could, or should, be built upon such foundations.

Chapter 2

The Hubble Universe

Memoirs of a Hapless Villain

> *The false dawn of universal expansion came from faulty data. Edwin Hubble was the first to realise it. "…it seems likely that red-shifts may not be due to an expanding Universe, and much of the speculation on the structure of the universe may require re-examination."*
>
> (Edwin Hubble, 1947).

◆ ◆ ◆ ◆ ◆ ❖ ◆ ◆ ◆ ◆ ◆

would imagine we all know what redshift is in principle, because we associate it with Doppler shift in sound, something we experience in our daily life. A screaming ambulance passes in the background, and we know without seeing it when it is moving away from us because the pitch of the siren drops. The departing ambulance stretches the sound waves and audibly lowers the sound frequency. In light waves, it's called Doppler redshift. The Doppler Effect—increasing wavelength caused by recessional velocity—is but one of many possible causes of a shift towards red in the spectrum of light.

Two observational pillars support contemporary cosmology: Hubble redshift and microwave background radiation. The first is used to describe systematic expansion of the Universe, and the second is put forward as a radiation image of the Universe as it was very soon after the Big Bang. Because redshift and radio noise are things that are seen to exist, much of the discussion on

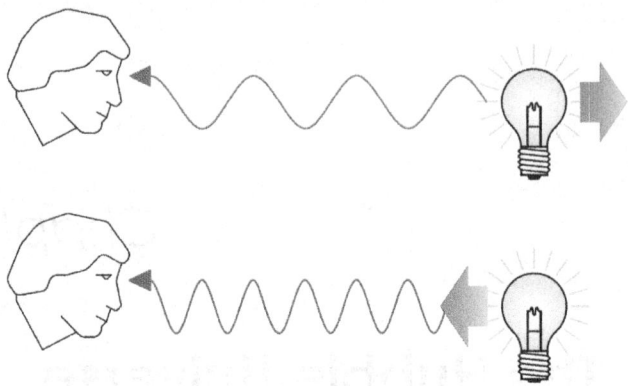

Figure 4: The Doppler Effect in light from a moving source. Recession (top) stretches the wave, moving spectral lines towards the red side of the spectrum, hence "redshift". (Thanks to Gregg Barlow for the diagram).

cosmology centres on one or both of them. Observational astronomers and astrophysicists using empiricism to derive their explanations of the cosmos would tend to concentrate on the first two tenets of cosmology, and in discussing the subject would lean towards redshift because the microwave background requires horrendous mathematical manipulation before it makes sense in the BBT context.

It is important that we carefully examine Hubble's own interpretation of this apparent redshift relationship. It started in 1924, with his observation of Cepheid variables in some of the so-called spiral nebulae. It was thereby established that these nebulae were in fact stellar systems in their own right, located far outside the Milky Way, and this constituted the advent of extragalactic astronomy.[*]

In 1929, Hubble discovered that for galaxies in his field of view, that is, fairly local, the fainter they appeared to be, the higher seemed the redshift. He might, he initially surmised, have found some progressive correlation between the distance to nearby galaxies (inferred by apparent brightness) and the overall redshift of the light coming from them. He was puzzled. Not so the cosmologists who heard of his discovery. They were delighted.

---

[*] The adjective galactic refers only to properties of the Milky Way. For offshore galaxies, astronomy has adopted the term galaxian.

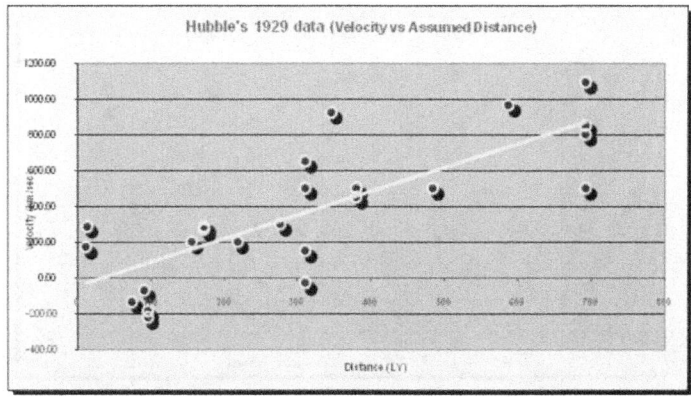

Figure 5: A plot of redshift (recessional velocity) versus estimated distance—the basis of the Hubble law. Plotting redshift against luminosity (figure six) paints a dramatically different picture.

For reasons that will soon become clear, they made the inference that it was caused by the Doppler Effect. In other words, they reiterated their belief that the galaxies were moving away from us, that the further they were, the faster they were going, and therefore that the rate of recession also told us how far away the object is. This is generally referred to as the Hubble Law. But it was observationally inconclusive, and Hubble himself remained deeply sceptical.

If we look at a plot of the 1929 Hubble data (Figure 5), mapping redshift against estimated distances, we are almost convinced. The trend line through the scatter could, with some kindness (and a deal of myopia) be considered an indication of something roughly linear. If we were looking for evidence, we would be inclined to recognise it here. However, these data are extrapolations of the observed quantities. A plot of the unaltered numbers (Figure 6) paints a dramatically different picture.

From the outset, however, data patterns were indistinct and tenuous. Hubble's original redshift data were described by Nobel Laureate Steven Weinberg as leaving him

> …perplexed how he (Hubble) could reach such a conclusion—galactic velocities seem almost uncorrelated with their distance, with only a mild tendency for velocity to increase with distance.[*]

---

[*] Steven Weinberg, *The First Three Minutes* (New York: Basic Books 1977).

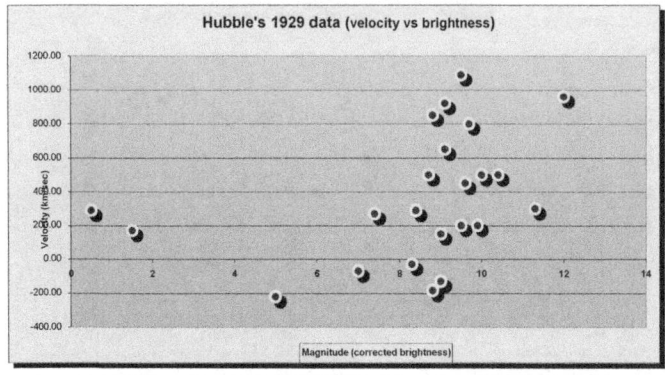

Figure 6: It looks like one big error bar! Hubble's original data for 23 galaxies, with velocity plotted against brightness. Compare this wide, almost vertical scatter with the later more linear plot he made against assumed distance (Figure five). (Note that magnitude is an inverse logarithmic scale – the higher the magnitude value, the fainter the object.)

To his eternal credit however, Edwin Hubble was one of those rare animals who could take his hat in his hand and admit to an error of judgement, even if that selfsame error was being acclaimed as an eternal truth by his peers. Hubble remained steadfastly unconvinced that the Doppler Effect correctly explained his observations and he was at pains to declare it quite emphatically. He clearly realised that no universal principle was revealed by his measurements, but he and later analysts missed a vital point: What was the redshift-brightness relationship *actually* telling us?

By 1934, it was agreed by all players that universal expansion was a remote phenomenon. Cosmologists, for once, were unanimous: The expansion hypothesis could apply only non-locally, that is to say, at the very least beyond the confines of the Local Group of galaxies. That effectively ruled out the applicability of Hubble's original sample. Whatever pattern he thought he might have seen in the spectra of those original galaxies, it was certainly not anything that indicated expansion.

There can be little doubt that Edwin Hubble and his colleagues were greatly dismayed by the revelation in 1934 that their galaxies were actually arranged in non-expanding space. But the implications were stark and undeniable—the only tangible observational data pattern raised in support of universal expansion

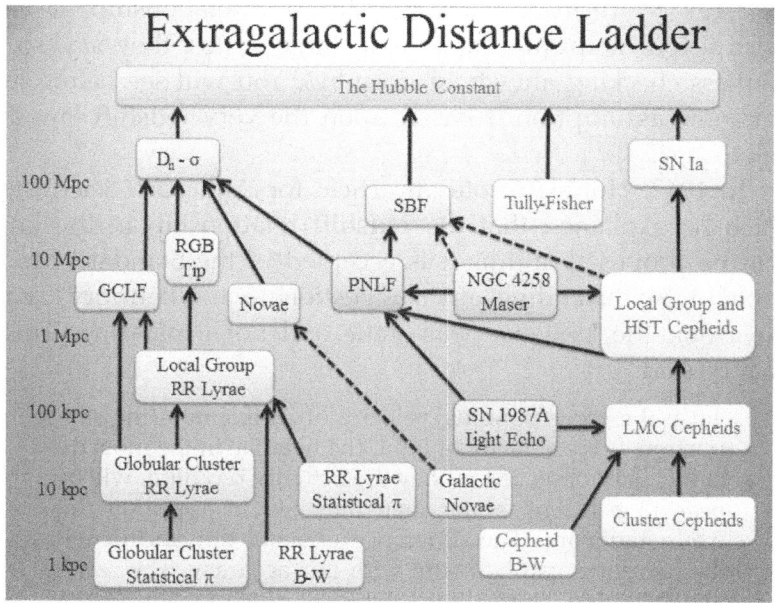

Figure 7: A graphical representation of extragalactic distance techniques. Note that all are related to the Hubble constant; none is independent of the Hubble law. (Image from Wikimedia Commons, used by kind permission of the author, Palmtree3000).

had just gone up in a puff of smoke. In 1935, he and Richard Tolman warned us,

> ...the possibility that red-shift may be due to some other cause, connected with the long time or distance involved in the passage of the light from the nebula to observer, should not be prematurely neglected.[*]

In a paper for the Astronomical Society of the Pacific in 1947 (some 20 years after his discovery), Hubble stated,

> ...it seems likely that red-shifts may not be due to an expanding Universe, and much of the speculation on the structure of the universe may require re-examination.

The reason is quite simple, but pervasive: The argument is circular, like a puppy chasing its tail. As we shall see in the chapter that follows, cosmological distance estimation is carried almost single-handedly by redshift. Look carefully at the studies

---

[*] Edwin Hubble and Richard C. Tolman, "Two methods of investigating the nature of nebular red-shift" *ApJ* **82** (1935), 302H; and Edwin Hubble, *PASP* **15** (1947), 773.

that seek to verify the redshift-distance relationship at large scales (expansion only happens at large scales, they say, so it's pointless checking anywhere close by). You will see before long that some assumption is based upon the very redshift law they seek to verify.

In 1942, Hubble wrote an article for American Scientist, in which he explained that the redshift relationship in the linear form he proposed and as it is accepted in the Standard Model, would notwithstanding any suggestion to the contrary clearly prove that the Universe was static and not expanding. This is what he said:

> Since the corresponding velocity of recession is the same fraction of the velocity of light, the nebulae in the most distant cluster observed, if they are actually receding, will appear 13 per cent fainter than they would appear if they were stationary. The difference is small but, fortunately, the measures can be made with fair accuracy. The results may be stated simply. If the nebulae are stationary, the law of red shifts is sensibly linear; red shifts are a constant multiple of distances. In other words, each unit of light path contributes the same amount of red shift. On the other hand, if the nebulae are receding, and the dimming factors are applied, the scale of distances is altered, and the law of red shifts is no longer linear.[*]

Without further ado, I could rest my case right there. If the authors of the Standard Model calibrate their expansion along an axis of linear redshift relationships—and they do, with great conviction—then they are telling us that their Universe does not expand. So why do I not rest my case and go home for tea? Well, *you* try telling them what Edwin Hubble said, see how far you get...

I'll give you an example. In 1983, the Infra Red Astronomical Satellite (IRAS) was put into orbit, and eventually returned over 350,000 catalogued infrared sources. The standard response to new parcels of data is to interrogate them carefully in case there might be hidden in the files some element of support for the favoured model. IRAS didn't escape this zeal, and soon, papers were being published celebrating confirmation of the Hubble Law.

---

[*] Edwin Hubble, "The Problem of the Expanding Universe" *American Scientist* **30** No. 2 (April 1942), pp. 110-111.

In 1996, eminent Princeton astrophysicist Michael Strauss joined forces with Daniel Koranyi of the Centre for Astrophysics, and together they threw their weight behind the effort. Their paper for the Astrophysical Journal was entitled "Testing the Hubble Law with the IRAS 1.2 Jy Redshift Survey."[*] Actually, their intent was not to test the Hubble Law for verity — that was assumed — but to refute an earlier study by Segal *et al.* which showed that the Hubble relationship was quadratic, not linear. Strauss was arguing that Segal was mistaken; the relationship is indeed conclusively linear.

Well, that's interesting. As we saw earlier, Hubble told us that if the law was linear, the Universe isn't expanding. Anyway, Strauss raised his arguments against Segal, and believes he is right. It would appear that he concludes the Hubble Law *is* linear after all, but that's not what I'm trying to point out. Strauss and Koranyi explain their method:

> This is done, in effect, by treating the entire galaxy luminosity function...as a distance indicator; equivalently, we compare flux density...with predictions under different redshift-distance cosmologies...

They test the redshift law using redshift law-derived data: The galaxy luminosity and flux density used in this exercise are redshift functions. That's blatantly circular. No test of the Hubble law completely independent of the same law has ever been done, to the best of my knowledge. I doubt it would even be possible, because the "law" is in any case artificial, manufactured synthetically from the data. That there could be doubt and ongoing debate over whether the law is linear or quadratic — a significant difference, it must be said — indicates how indistinct the supposed data patterns are.

And so it remained. In his own heart, Edwin Hubble's famous revelation perished on the horns of a fearsome dilemma. At first, he had idealistically developed a cosmology based upon the exciting but untested theoretical supposition that the entire Universe was expanding away from a beginning in time. Then came that tantalising whisper, a preconceived thread that he read into the relationship between the apparent luminosity of galaxies and their redshift.

---

[*] Daniel Koranyi and Michael Strauss, "Testing the Hubble Law with the IRAS 1.2 Jy Redshift Survey" *ApJ* **477**, 36.

In the words of Steven Weinberg, here was "the answer he wanted to get". It was a miscue that made his name one of the most revered in cosmology. Unfortunately, Hubble could not sustain that point of view. He was soon forced to abandon his initial position in its entirety, but it was too late; the damage had been done.

Nevertheless, he demonstrated the courage of his convictions and a gratifying grasp of reality. As a consequence of his observations, Edwin Hubble developed a description of the cosmos that was both *infinite* and *static*, the diametric opposite of both his original model and the current Standard Model of Cosmology. He was certainly fearlessly honest, and did not hesitate to state this opinion in writing and in lectures.

Hubble and Tolman were driven by their nagging scepticism to suggest a test for the redshift/velocity relationship, the now well-known number effect. The following quotes are from Hubble's own writing (in his 1936 book The Realm of the Nebulae, commencing page 193):

> Apparent luminosity is measured by the rate at which the quanta reach the observer, together with the energy in the quanta...The rate of arrival (*i.e.*, the number of quanta reaching the observer per second) is reduced if the nebulae are receding from the observer, but not otherwise. This phenomenon, known as the 'number effect', should in principle provide a crucial test of the interpretation of redshifts as velocity-shifts.[*]

And indeed it has. To the best of my knowledge, no application of this test has succeeded in verifying the velocity interpretation, despite numerous attempts. On the other hand, at least two published studies (Andrews, 2006,[†] and Lerner, 2006[‡]) have shown quite conclusively that the number effect in fact falsifies

---

[*] Cited in A.K.T. Assis, M.C.D. Neves, and D.S.L. Soares, "Hubble's Cosmology: From a Finite Expanding Universe to a Static Endless Universe" *arxiv*: astro-ph/0806.4481.

[†] T.B. Andrews, "Falsification of the expanding universe" in: E.J. Lerner and J. B. Almeida, editors, *Proceedings of the 1st Crisis in Cosmology Conference: CCC1*, American Institute of Physics (2006).

[‡] E.J. Lerner, "Evidence for a Non-Expanding Universe: Surface Data from the HUDF" *arXiv*: astro-ph/0509611, in: E.J. Lerner and J. B. Almeida, editors, *Proceedings of the 1st Crisis in Cosmology Conference: CCC1*, American Institute of Physics (2006).

the expanding Universe model, and cosmological redshift is thereby released from its velocity cage to be freely interpreted without bias.

Not quite. That would be wonderful were it true, but astronomy icon Alan Sandage rained on my parade. At least, he made a stout effort to do so. In June, 2009, he submitted a paper to the Astrophysical Journal that threatened to prove me wrong. It's a 51-page doorstopper with an equally massive title: "The Tolman Surface Brightness Test for the Reality of the Expansion. V. Provenance of the Test and a New Representation of the Data for Three Remote HST Galaxy Clusters."[*]

My heart sank when I saw it, but I should have known better. It's not arrogance but rather quiet confidence that lets me sleep soundly between observations. As a physicist used to dealing with real things, I know that the expansion paradigm is more than extraordinary, far beyond unlikely, just hopeless wishful thinking. I should be very surprised if any observation or experiment can be contrived to unambiguously support it. Even someone with all the vast experience and expertise of Alan Sandage falls prey to the comfort of serving the orthodoxy.

I think we will find that fluctuations in the energy levels of light will be an effect resulting from a cocktail of causes *because space is not empty* (see chapter four). We can consequently state with certainty that some weariness will result as light fights its way across the Universe, quite irrespective of whether there is universal expansion or not. As deeply learned polymath Ari Brynjolfsson has pointed out, Sandage's procedures are meticulous but his results are ambiguous. There could be many causes of the observed effect, and certainly, the mechanisms that I list in chapter four all employ good physics, something that cannot be said for the creation of spacetime between gravitationally bound objects in the Lambda-Cold Dark Matter Model.

As we plough through the complex maze of geometric factors like Petrosian radii and Sérsic exponents—it is illuminating to note that Sandage prefaces his analysis with the old Chinese saying, "Details sire the complications"—it becomes clear that an essential constraint on the Tolman test has been violated: It

---

[*] Alan Sandage, "The Tolman Surface Brightness Test for the Reality of the Expansion. V. Provenance of the Test and a New Representation of the Data for Three Remote HST Galaxy Clusters" *arxiv*: astro-ph/0905.3199.

should be completely independent of any model, and this study is not. Louis Marmet pointed out in private correspondence (referring to Sandage's earlier Tolman studies),

> Sandage explains that the Tolman test should be independent of the cosmology, but a calculation of the absolute magnitude of the galaxies is required to be able to classify them ... So, although the surface brightness is an absolute quantity, the identification of the galaxies (and therefore their absolute luminosities and diameters) is dependent on the cosmology.

In simple terms, surface brightness is flux density (luminosity divided by surface area) and in order to determine area, we first need actual, not angular, diameter. Of course, to translate angular diameter to real diameter we have to know how far away the object is. Remoteness cannot be measured; it is given by inference from the model (redshift and the Hubble constant). It is clear that Sandage's reasoning is model-dependant and therefore circular, termed *circulus in probando* (using the conclusion as an argument).

Alan Sandage, with the greatest respect, looks at the data through the opportunistic window provided by contemporary cosmology, and uses whatever technique is available in the library of physics and mathematics to get the jigsaw puzzle pieces into the right shape to fit the gap in the picture. Fred Hoyle once described this approach to cosmology as "complicating everything to the point of incomprehensibility." Sandage doesn't make the crucial breakthrough after all. Nice try though...

Hubble carried the static, endless model with him to the end. Hoyle, Burbidge, and Narlikar, in their book A Different Approach to Cosmology, recount Hubble's concluding uncertainty:

> In his last discussion of the observations, Hubble in the George Darwin lecture at the Royal Astronomical Society in 1953, a few months before he died, gave the first results obtained using the 200-inch telescope...Sandage has pointed out that using the 'no recession factor' (meaning no correction for the number effect), Hubble was still doubtful if the expansion was real.[*]

Sadly, it seems that if the discoverer of cosmological redshift was doubtful, the same could not be said of certain interested by-

---

[*] Fred Hoyle, Geoffrey Burbidge, and Jayant Narlikar, *A Different Approach to Cosmology* (Cambridge: Cambridge University Press, 2000).

Figure 8: Edwin Hubble at the 100" Hooker telescope at Mt Wilson, circa 1928.

standers. The eager cosmologists waiting in the wings were, true to form, not in the slightest doubt whatsoever that the whole Universe was flying apart and that Hubble had proved it. What a great shame that our memory of Edwin Hubble is almost indelibly tainted by their undue haste to make a point.

Hubble's discovery (pardon me for using a misnomer) seemed to have made everyone happy except Albert Einstein (and Hubble himself, of course), but eventually even Einstein grudgingly conceded, when faced with the solutions to his equations by Russian mathematician Alexander Friedmann, that expansion was an inevitable consequence of General Relativity. We need to understand the implications of this. Despite pressing anomalies, progressive expansion, with every point of view the centre of its own carbon-copy universe, came to be entrenched in cosmology and astrophysics. Once the Hubble Law interpretation of redshift was accepted as fact, astronomy geared itself to using it as a tool to map the universe. It says the higher the redshift, the further the object, and therefore the higher its recessional velocity. All well and good, except for one thing: Doppler shift just didn't work. Recessional velocity would soon reach the speed of light, then exceed it, and that simply wouldn't do. In fact, the use of the redshift value z tacitly admits this. We are going to be dealing with z extensively in our investigation, so let's define it. Essentially, it represents a ratio between the recessional velocity of the object being studied (v) and the speed of light (c), thus

$$z = \frac{v}{c}$$

Therefore, by simple deduction, if the value of z is greater than 1, then v is greater than c. That means, obviously, that the galaxy in our crosshairs is receding faster than the speed of light, and z represents the factor by which it does so. So, if we say that for an object in our view, the redshift value z is 0.5, we imply that it is receding from us at half the speed of light, and if $z = 5.3$, then it flies away at more than 5 times the speed of light.[*] Let us not be without some empathy here; this is a very serious problem indeed, and it took a devil of an effort to quell. The theoreticians behind BBT had to put their thinking caps on again. The prospects were bleak, but defeat was simply not an option. Then came the eureka moment. Aha! The answer, they told a relieved astrophysical world, was this: Although the galaxies weren't actually moving apart, the space between them was expanding. That stretched the light waves, and dilated time itself, without causing the measurable distance between the galaxies to increase.

Before we move on to other pastures and fresh contemplation, we should discuss the "subsequent work" so often alluded to but seldom decently identified in articles and papers about the Hubble Law. Surely there have been more recent tests, using modern equipment? Indeed there have; several in fact. All those that I have seen are unanimous in their support for the Hubble Law and concomitant expansion. Did the later redshift-luminosity data succeed where Hubble's original effort had failed? That question haunted me. The way to check it out would come to me quite unexpectedly on a dark and windy night in the mountains. Paul Jackson, a retired physics professor and trusted confidant, lives in an intriguing, charmingly Heath-Robinson, self-built home on the inland slope of KwaZulu-Natal's Karkloof range. From time to time I visit him there, usually to take advantage of some fresh mountain air, good farm cooking, and solid advice.

---

[*] Alert readers might be inclined to point out that it is more correctly expressed as $cz = \Delta\lambda/\lambda$, but it boils down to the same thing.

The night in question was Dickensian in its misery. The freezing wind howled through the whipping pines behind us, and anyone outside must have been convinced that ice, not fire, would signal Armageddon. Inside though, I was as snug as a bug in a rug, quite unaware of the impending epiphany. My bedroom doubled as Paul's study, and I was delighted by the prospect of exploring his pregnant bookcase. I pulled a large, dog-eared book from the shelf and settled down to read.

One of the standard texts in the field is the definitive volume The Principles of Physical Cosmology by eminent Princeton physicist P. J. E. "Jim" Peebles.[*] The context of what follows will be taken from his concise summary of the expansion concept on page 71:

> The expansion of the universe means that the proper physical distance between a well-separated pair of galaxies is increasing with time, that is, the galaxies are receding from each other. A gravitationally bound system such as the Local Group is not expanding ... the homogeneous expansion law refers to galaxies far enough apart for these local irregularities to be ignored.

There you have it, in a nutshell, from the pen of one of the most revered spokesmen of consensus cosmology. Expansion, and indeed any consistent sign of it, can only exist at extremely great but apparently indeterminate distances.

Like the persistent whine of a determined and hungry mosquito, the notion of non-locality hovered subliminally in the recesses of my mind, and as we shall soon see, improperly tinted my spectacles on this occasion. On page 50 of that book, Figure 3.13 is a graphical representation of the correlation in a sample of elliptical galaxies of their velocity dispersion (represented by σ, the Greek letter *sigma*) with their apparent luminosity.[†] There is what appears to be a linear trend through the scatter of data points in the plot, so for the sake of argument, let's assume that there is a real trend in the data. Theory relates velocity dispersion to cluster mass, and mass in a body of incandescent stars is pro-

---

[*] P.J.E. Peebles, *The Principles of Physical Cosmology* (Princeton: Princeton University Press, 1993).

[†] Velocity dispersion is the spread of velocities of stars or galaxies in a more or less spherical cluster. It is estimated from the radial velocities of selected component objects in the group, and once established can give the cluster mass by means of the virial theorem.

portional to intrinsic brightness (because, simply put, more mass means more stars, and therefore more light). What does this actually tell us? Certainly not what I thought at the time, and somewhat less than Peebles implies.

My weariness must have blurred my concentration somewhat, because (as Paul later pointed out) I mistakenly took the diagram to represent a direct extrapolation of the relationship Hubble tried to establish in 1929 (redshift versus measured brightness of galaxies), whereas Peebles plots the velocity dispersion of stars within galaxies without invoking redshift of the galaxies themselves. It doesn't particularly worry me that I made a mistake; I often do, and gladly admit my error as soon as it is revealed to me. In this case, it was the principle involved that pitched a curve ball at the science I was tracking, and gave me a positive clue to the Achilles' heel of redshift cosmology.

I consider it vital that we take due cognisance of a pervading habit in any zealous search for observational evidence. This treatment of observationally acquired data sets has haunted relativistic cosmology since its inception: Commencing with the eclipse data reported by Arthur Eddington in 1919 and punctuating the development of Big Bang Theory all the way through to the latest claims being made in the first decade of the 21st century, evidence is somehow found in observational measurements that either does not meaningfully exist in the pristine data, or if a pattern *is* found, does not refer to or in any way validate the preferred theoretical model.[*] Objectively inconclusive results are given meaning that closer analysis reveals to be pointing in another direction completely. It's a dangerous game. Like a cornered dog, synthetic evidence can bite you, and in the case of establishing a trend of luminosity versus redshift, it bit. What I needed to do was find the wound. I did find it, some time after my return from the Jacksons, and further careful inspection of my own copy of The Principles of Physical Cosmology provided the crucial and long-sought breakthrough.

What struck a chord for me was that the galaxies in Peebles' sample are ellipticals from the Virgo and Coma clusters. We all know that the postulated expansion of space does not occur lo-

---

[*] In my opinion, it started well before Eddington's blatantly censored Principe and Sobral eclipse data. The Michelson-Morley experiment of 1887 is a case in point. However, we cannot afford to be distracted by peripheral arguments right now.

cally, and "local" includes the Virgo cluster and almost certainly also the Coma cluster. With unsubstantiated optimism, the standard theory alludes to a threshold for expansion at around 100 Mpc from the Earth, meaning that for the first 350 million light years or so, space does not expand. Any perceived pattern in these data *cannot* indicate expansion, in terms of Big Bang Theory. This would be an utter train smash for the Hubble law if only I could find proof in the form of a published data table or graph.

It wasn't hard. It's right there in black and white on page 86 of Peebles' book. Figure 5.4 bears the caption, "Test of Hubble's law using Tully-Fisher distances."[*] Before we continue, I wish to acknowledge Jim Peebles' self-deprecating honesty in the statement, "The distances in figure 5.4 are expressed in megaparsecs, but this is based on the still somewhat controversial calibration of the absolute magnitude-δv_{21} relation."[†] We shall be discussing this controversial uncertainty in the next chapter.

The plot in the diagram shows the Hubble relationship established in the supposed redshift-distance correlation for a sample of galaxies in the vicinity of an object popularly identified as the Great Attractor. Although it has never been seen (it would in any event be obscured by the Milky Way's disk), it has been invoked to explain the peculiar streaming motion of galaxies in the neighbourhood. A team led by Lyndon-Bell discovered in 1988 that peculiar velocities in this region are puzzlingly large, around 600 km sec^{-1} for the entire Local Group, and this could only be explained by the presence of an extremely massive object somewhere in the direction they were headed (Aside: this also caused a bad headache elsewhere in consensus cosmology, because the anisotropy—a local effect—shows up persistently in the CMBR, which of course is expressly forbidden by underlying theory).

The crucial significance of this geographical location is twofold: Firstly, it is local (all galaxies on the plot are <100 Mpc); and secondly, the presence in this locale of a structure massive enough to divert entire clusters of galaxies from the mooted Hubble flow is in defiance of the Cosmological Principle, and therefore rules out Hubble expansion in the region being ob-

---

[*] The Tully-Fisher relation is a robust correlation between internal rotational velocity in spiral galaxies (a function of stellar abundance) and their intrinsic luminosity. See chapter 5 for further discussion.

[†] The term δv_{21} refers to the width of the atomic hydrogen 21cm radio line from the galaxy disk, a standard measure of rotation.

served. Despite the fact that all parties to the debate would agree that the galaxies represented in the graph occupy a volume of space that is definitely not expanding, Peebles is quite clear in his conclusion about this particular plot: "We see that, even with the anomaly in the direction of Centaurus, Hubble's law is quite a good description of the redshift-distance relation."[*]

There you have it. Bingo! The Hubble law shows up in non-expanding space, and would therefore manifest in a static Universe. Hubble's 1929 discovery and all the subsequent developments upon it are clearly invalid as indicators of universal expansion. As I perused further in The Principles of Physical Cosmology, I quickly saw that there is an abundance of such observational evidence refuting the notion of redshift-verified expansion, but of course I need only one substantive example to make my point.

At the risk of labouring the point, here's the principle: Any correlation in observational data, perceived or real, between redshift and brightness cannot be taken to indicate expansion if it is also seen in static space. In fact, by their own logic, Standard Model theorists should concede that observationally, a linear relationship between the redshift of local galaxies and their apparent luminosities indicates quite the opposite: A static universe, not an expanding one.

Before we proceed, it would enhance our perspective considerably if we clearly grasp that on the flimsiest evidence, universal expansion was readily—hastily even—incorporated into the Standard Model of Cosmology. This was despite conceptual paradoxes so great that they have not yet, after nearly 80 years of intense speculation, been resolved. We have seen that the original data announced by Edwin Hubble in 1929 were within a few years found to be meaningless. There is no expansion, theoretical or otherwise, in the space between local galaxies, so whatever pattern it was that Hubble and the eager patriarchs of Big Bang Theory thought they had found in the numbers, it was certainly not related to the systematic recession of those galaxies. Yet, despite the fact that the data were never replaced, and no further evidence of systematic redshift ever found, the illusion was taken to heart with such determination that it became law! If we did not understand the nature of dogma before, now we do.

---

[*] P.J.E. Peebles, *op cit.*

Edwin Hubble has been badly served by history, and it's a shame. Despite his own rather colourful account of his early years, he matured into a scientist of great objectivity and courage. Integrity he had in abundance. Who else would have stood before flamboyant global adulation and declared his personal doubt that he had in fact discovered what he was credited for? For that alone, he should be carried shoulder-high in our memories; he would have been the most deserving Nobel prize-winner in the history of that award—not for the Hubble law, but for denying it. As hard as it is to accept, and no matter how passionately we love the Standard Model, if we are capable of being brutally honest with ourselves then we are left with no choice. It is simply unavoidable upon sober examination of what actually transpired in cosmology in the 1930s that the lauded Hubble law of recessional velocities, backbone of deep sky astrophysics, is a myth.

So, before we depart for the next chapter to put an historical perspective on how we calculate distances, let me ask you this question: When you consider all these imposed shapes, forces, fields, and velocities, all of which are invented to suit, doesn't your gut tell you there's something horribly wrong with the theory and practice of an expanding Universe?

Discussion: There is no real evidence for expansion. It is unsupported by observation and actually contradicted by it. Edwin Hubble did not discover the Hubble Law; it was born in a fuzzy trend in faulty data, and then ideologically insulated from conflicting observation. The original data indicating expansion were found to be specious, abandoned, and never replaced. It came from nowhere and, observationally, it seems it's going nowhere.

Chapter 3

The Distance Ladder

Where push comes to shove

> *How do we calculate the remoteness of distant cosmological objects, and can we be sure of the results? The answer in both cases is almost always, we cannot. "Each step on the distance ladder introduces further uncertainty. Would it not be better to use primary indicators to calculate galaxy distances, and thus remove the need for the treacherous distance ladder?"*
>
> (Stephen Webb in *Measuring the Universe*).

✦ ✦ ✦ ✦ ✦ ❖ ✦ ✦ ✦ ✦ ✦

" 𝒢 alaxy redshifts out to $z \sim 1.4$ can be obtained from optical spectra. At higher redshifts…one enters the so-called 'redshift desert' … One cannot begin to study the evolution of galaxies unless one has some idea of the redshift at which they lie."[*]

In the quotation above, a team of respected scientists from leading universities play their part in propagating and entrenching the idea that deflection towards red in the spectra of cosmological objects indicates, in some sense or another, the remoteness of the object. Implicit also in the term "redshift desert" is the ad-

---

[*] You will recall from chapter two that the redshift value z is the ratio of recessional velocity to the speed of light, thus $z = v/c$. Quote is from J. E. Gunn *et al.*, "Understanding the Astrophysics of Galaxy Evolution: the role of spectroscopic surveys in the next decade" *arXiv*: astro-ph/0903.3404.

mission that the idea doesn't work very well at all, or reach to any useful extent in space.

Redshift does not indicate remoteness. I'd like now, for once and for all, to put a lid on it. You see, that's exactly the problem. Galaxies do not lie *at* redshifts. It is, in a manner of speaking, the *redshifts* that lie!

Quantifying depth or radial distance from our point of observation is arguably the most daunting and pressing problem facing astrophysics. We are defeated by our inability to properly check measurements made with theoretically-derived techniques, and although that in itself is not a fatal error, refusal to admit to our limitations very nearly is.

No one is swaggering here. We're all in the same boat. There is a gratifying degree of circumspection evident amongst those engaged in tackling this problem and I wish it would spill over to other areas of cosmology, or even more pertinently, sustain its influence all the way to the frontier of cosmological distance measurement. Ongoing creativity tempered by empirical auditing is admirable right up to the point where redshift takes over. Then, with a huge sigh of relief, astronomers hand over to the grand redeemer of cosmological distance: Redshift. All the uncertainty inherent—and proper—in dealing with large-scale distance estimation almost magically disappears. I will argue that when all is said and done, redshift is not the anticipated redeemer at all; it is actually the villain of the piece.

The spatial arrangements in three dimensions of objects we see on the sky, and the quantified relationship between them, sets the ground and marks the field of play. If we cannot be sure how far away something is, it casts doubt upon many other properties woven into the overall picture. Intrinsic brightness, also called absolute luminosity, is just one example of many, albeit, as we shall see, one of the most crucial.

There is an established and growing set of techniques applied to the calculation of astronomical distances, working their way outwards from the Earth as if the Universe were arranged as the layers of an onion. This hierarchy of formulae is known as the Cosmological Distance Ladder, and it progresses radially outwards from our point of observation, each measurement regime handing over to the next rung on the ladder as it falters and dies in the frozen uncertainty of deep space. The reliability of each succeeding step depends wholly upon the proven ability of the one before it.

In astronomy, we measure both angular distance and, with irrepressible optimism, radial distance. The former is merely a ratio, measured in degrees of arc. In other words, what proportion of a full circle is represented by the apparent separation of two objects on the sky? To convert that apparent relationship to a real distance expressed in units of linear measure, we would need to know how far away both objects are. We cannot get a real idea of the 3-dimensional structure of the universe without a definite derivation of radial distance along line-of-sight. That, my friends, is where we slam unceremoniously into a frustratingly opaque wall of ignorance. How on Earth do we measure real distances at cosmological scales?

Alan Sandage rues the glaring absence of empirical backup to theoretical measurement: "Barring such a test (none has yet been successful), a direct verification that the redshift is due to a true expansion of the geometrical manifold would be most helpful, but again such a demonstration is not quite available yet." Observational tests based on linear measurement are patently weak. Alan Sandage again:

> The angular size—redshift relation should be the most direct way to sample the geometry. The theory of this test leads to the surface brightness $\sim (1+z)^{-4}$ relation, which must be valid if the expansion is real ... The search for a suitable measure of a metric size is the present stumbling block, one yet to be adequately dislodged, but progress has been made.

I suppose we could live with the optimism expressed in the word "progress" here.[*]

As usual, this will involve a bit of history. We should look back in time to remind ourselves by what means distances to remote objects—that is, objects beyond the grasp of direct, physical measurement—were derived. Having established that, we shall examine the relationship between the methods used so that we can determine whether or not the formulae used in this endeavour have been checked and verified to acceptable standards of science. Is there a useful overlap between rungs on the ladder, so

---

[*] Alan Sandage, "Observational Tests of World Models" *Annu. Rev. Astron. Astrophys.* **26** (1988), 561-630.

that we can be confident as we venture nervously outwards? Let us try to expose unseemly haste, if it exists. Unfortunately for us, we need to make a diversion into theory again before we can get on with the real meat on these mental bones. What sort of space are we attempting to measure in, and does it really matter?[*]

We live in a 3-dimensional Universe. We can argue about dimensions and geometry until we are blue in the face, it will help us not a whit. What we need is some kind of experimental demonstration, of a kind that by elimination indicates what configuration of space we are dealing with in the real world. Something that tests mathematics against itself is patently no good; it must be a test in physical reality with consistently measurable results.

Is such a material demonstration of mathematical abstraction at all possible? I have argued the point with some highly regarded mathematical theorists who are adamant they can and have done it successfully, even if it was to their own satisfaction only. They claim that their experiments demonstrate to observers what kind, form, and shape of space they stand in. With respect, I strongly doubt it. How are we to test mathematical representations of *space* against reality? Any experiment we perform is naturally taking place in space as it actually exists, regardless of how we describe it. Is it practicable to perform a simple comparison between these particular model-dependent theoretical descriptions of a physical object with what is measurably real on the laboratory table? To my mind, it would appear not, at least not as long as we think in geometrical abstractions. It seems to me that mathematical description is proof against this type of approach, by virtue of the fact that it eliminates the validity of objective experience.

Figure 9 is artist Gregg Barlow's rendition of something called a "Klein bottle." (It is also on the cover of this book.) A Klein bottle is often described as having neither inside nor outside, and if you follow the contortions of the structure, that certainly appears to be the case. Actually, the Klein bottle, first conceived by mathematician Felix Klein, was originally called by him *Kleinsche fläche*, which translates as "Klein's surface." This was incorrectly read by English translators as *flasche*, meaning *bottle*,

---

[*] Chapter eight examines geometrical curvature specifically and in more detail.

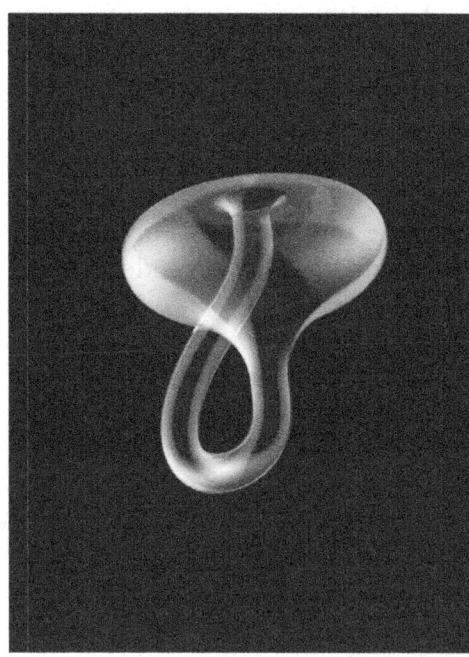

Figure 9: A geometric shape, originally thought to be abstract, called a Klein bottle. Does it verify 3-D Euclidean space? (Diagram by Gregg Barlow).

and the name stuck. Geometrically, it is a 2-dimensional surface rather than a 3-dimensional object.

A Klein bottle doesn't demonstrate anything useful as far as I can see, except that in geometry a surface can be considered in isolation from a third spatial dimension, whereas in our mental conceptualisation, and of course in physical reality, a surface is always embedded in or appended to something with depth as well as breadth and height. We *think* 3-dimensionally, and things *exist* 3-dimensionally. Thank heavens for that—there would be more than usual chaos if this coherence weren't the case. Here is the point I wish to make: The idea of a materially isolated surface is not capable of physical realisation.

What we need to do is emulate what nature has done: Create real objects. Then we can look at them to see what that reveals about the shape of space to us. Perhaps we could test the notion of space dimensionality by comparing models, as follows: Using clay, model the globe of the Earth. Draw a line on the surface of the sphere. Now tell me, there on the physical model before you, is the line straight or is it curved? We could *imagine* that it's straight in some mental universe, but here in the laboratory, under the glare of harsh reality, it is without a shadow of a doubt *curved*. Do we agree?

In Euclidean space, a straight line is without guile. It is properly straight by any objective test. In Riemann space, this is not the case. A line drawn to connect two points on a curved surface by the shortest route can be considered "straight" in the sense that it is the shortest line connecting the two points of interest. A straight line can be curved in the sense that it is a geodesic (shortest route) on a curved surface. So, I reason, let's see if the space surrounding us is Euclidean or Riemannian. Draw a straight line on a rigid surface. Test it. Is it really the shortest distance between its endpoints, or does it take a roundabout route? If it is straight, then Riemann space is eliminated and Euclidean space confirmed. The physical existence of Klein bottles in gift shops demonstrates Euclidean space to my satisfaction, as does every other material object I've ever come across.

I am told that this argument is so weak, it's almost embarrassing. Maybe so, but that's exactly how I feel about the tests conducted by my critics, and in fact, about the entire issue of what things look like in what type of space. It's irrelevant. Here's the bottom line: We cannot agree on the validity of empirical evidence for the type of space we live in, and until we have it, let's just use space as it manifests around us, no matter what name we call it. If we assume that it is the same for all of us then there is no need to factor it into our calculations.

This is important to our investigation, and although it may offend some geometers I know, we are going to remove space-model theoretical manipulations from our equations. We will take unprejudiced measurements and note the quantities returned. Furthermore, we assume Newton's third rule of reasoning: What's good right here is also good over there, until measurement indicates otherwise. We shall simply take space at face value, a parameter constraining all observations equally, and map it along three orthogonal spatial axes of unvarying calibration. It's simple and it works perfectly well in our laboratory.

Right, back to the task at hand. How do we determine distance in familiar space? We are blithely building a 3-dimensional picture of the world around us, based upon the dubious assumption that we can measure distance reliably on very large scales. We cannot. It is mediocre guesswork at best. The further out we go, the more we tend to clutch at straws.

Please understand that I am not being disparaging; merely realistic. We are faced with intensely daunting challenges in astrophysics, none of them more so than the general problem of measurement at great remoteness. We establish a technique, from theory or experiment, which appears to hint at distance. We apply it to the environment, look for consistent patterns, and if we are lucky, test it against proven measurements. The latter is a luxury, for, as we shall see, a second opinion confirming the diagnosis rarely presents itself to the ailing cosmological distance ladder. We are thus dependent upon measurement techniques that have not been properly validated, and which guess at distances with all the finesse of a three-legged porcupine.

There is, of course, the further (I believe, entirely unnecessary) complication of expanding space and the array of sometimes quirky effects it might produce in the light signals we deal with in astrophysics. Not only are cosmological objects in this model constantly moving away from us—and consequently at greater and greater distances with each passing moment—but also, in ways we do not fully comprehend, having the optical properties of their images variously distorted by intermediary objects and by the redshift process itself.

At an elementary level, this affects a crucial component of the distance ladder: Luminosity. Implied Doppler-redshift leads to a standard adjustment to perceived luminosity called the K-correction, something that is done as a matter of course in the process of celestial analysis. But should it be automatically applied? The thinking behind it is that spectral redshift causes observed (apparent) magnitude to differ between the zero-redshift rest frame of the source and non-zero redshift at the observer. This is because the photon flux (energy) diminishes with redshift, and we consequently see the object somewhat less brightly than we would have had it not been running away. We therefore multiply observed magnitudes by $2.5\log(1+z)$ to brighten them up a bit on our data tables, and everyone's happy.

So what's the problem? It is this: Cosmological calculations are being performed on the assumptions of a) expansion, b) Doppler redshift, and c) local $z = 0$. Notwithstanding that b) has long since been abandoned by the Standard Model, this is obviously model-dependent cosmology, and in effect, measurements are being doctored to suit the *a priori* preferred theoretical framework. We are not considering the observational data objectively, and that is bad science. I have yet to hear of a cosmologist who disagrees with the principle that cosmological calculations should be undertaken independently of any model, yet such data censorship is rife in the field.

Another example is the Integrated Sachs-Wolfe Effect (ISW), devised in an attempt to explain the uncooperative power spectrum of the microwave background. It says that light covering the vast journey from the primordial Universe to reach Earth in the present time has interacted rather peculiarly with massive objects along the way. These objects are referred to as gravitational wells. They are supposedly transparent, allowing light waves to pass through without deflection.

However, a mooted differential in the rate at which space expands in certain parts of the Universe imprints a signature on the light signal, and, we are assured, this can be seen and read if we look at the Cosmic Microwave Background Radiation carefully enough. Certainly, there are patterns in the CMBR, but caution is needed. We must bear two things in mind: Firstly, many of these effects are visible only because they have been artificially enhanced to suit the parameters of a particular investigation or investigator; secondly and far more importantly, all of the reported CMBR patterns would be equally apparent in radiation that came from natural astrophysical objects. In other words, the signs in the power spectrum are clearly ambiguous.

The question we really ought to ask ourselves is, do they come from what the theory suggests, or is there an altogether simpler explanation? The ISW is terribly complicated, and entwines itself with another linear effect called the Rees-Sciama Effect, which bounces around with yet another effect called Sunyaev-Zeldovitch, but I shall be brief. It's all too easy to become distracted by these things.

The Integrated Sachs-Wolf effect occurs when, in expanding space, the wave falls into, then climbs out of, gravitational wells. As it falls towards the centre of gravity in the object, the wavelength is compressed by gravity. On the other side, the expansion stretches the wave again, and because more expansion takes place on the exit than in the entrance (as a function of elapsed time), the light ends up more redshifted than it ought to be. This does of course ignore the tenet of expansion theory that excludes the inside of gravitationally bound systems from expansion, a stumbling block yet to be overcome by theorists, as far as I know.

Pertinent to our discussion here is the suggestion that red-shift excess is caused by expansion along the radial axis within intervening structures during the travel-time of light. However, unless we know precisely how many gravitational wells are involved, how big each is precisely, and where they are situated on the time-line, we haven't a snowball's hope of making a calculated correction to the spectral deflection to leave only Doppler shift. Without a doubt, having to allow for expansion along line-of-sight makes the problem of cosmological distance immeasurably more daunting.

Against that background, we see a somewhat naïve tendency to derive distances to cosmologically remote objects by using the Hubble law. The sheer complexity of calculating radial distances at these scales makes redshift appear to be an oasis of simplicity. It seems elementary enough: *Redshift*, via the Doppler Effect, gives *velocity*, and *velocity*, in a systematically expanding, smooth universe, gives *distance*. Divide the redshift value by the Hubble constant and distance is obtained.

> The Doppler interpretation (that is, recessional velocity v at radial distance r as a function of z) gives $v = cz$ (where c is merely a constant to relate units of measure*) and $v = H_0 r$ thus, $r = cz / H_0$

Therefore, the distance r to an object is the Doppler velocity divided by the Hubble constant. There are pitfalls however, some of which are in my view insurmountable. It calls our assumptions into question, and a thorough search for the data underpinning the *redshift-velocity-distance* hypothesis comes up empty handed. The fact of the matter is that quantifying any aspect of an expanding Universe is entirely model-dependent, to the extent that the "constant" legitimising the equations of distance or velocity is—astoundingly—a *variable!* The Hubble constant H_0 is used in every equation of cosmological distance, yet it runs with the hares and hunts with the hounds.

We should spend some time and effort getting to grips with the Hubble constant. I must caution you, though; it's not nearly as straightforward as it pretends to be. If I want to measure the distance from my nose to the computer screen before me, there are no complications. Only one space, one distance, one geometrical

---

* In fact, it can validly be argued that c, in all matters of Relativity, is not the speed of light at all, but a constant to render results dimensionless.

framework, and no variables or fudge factors. Before we're through with H_0, let me warn you right now, we will have had to deal with five different types of distance, three distinct renditions of space, three types of mass-energy (two of which are undetectable), three opposing geometrical models, two fundamentally new, unprecedented fields of physics, a varying fundamental constant, and considerable uncertainty. But that's all. After that, we're done. Relax.

For decades, cosmologists were happy that expansion of the universe was unspectacular in the sense that by normal physics, the explosion would have supplied the initial kinetic energy to drive expansion, and the rest would have been momentum succumbing to the overall mass attraction of everything that was. It's quite straightforward, like tossing a ball into the air. It will soon reach parity between momentum and gravity, and thence return to Earth. No rocket science needed.

No such luck with the cosmos, I'm afraid.

I suppose we choose a preferred model by some arbitrary means, perhaps even by common sense. It would be pleasing to discover that we had used logic and deduction from first principles to select a model that best fits our observations and experience, but that's not how it gets done in astronomy; quite the reverse.

At the outset, the decision needs to be made: Do we in reality occupy an Einstein-de Sitter universe? Let us remind ourselves what that might be. Albert Einstein and Willem de Sitter both developed cosmological models in 1917. Both were solutions to the equations of General Relativity, yet they could hardly have been more different. You see, the universe proposed by de Sitter was entirely and absolutely empty! It was just space and space alone, and it expanded. No, don't expect me to tell you how it works, because it's much too late to be puzzling over things like that. But I'm glad you asked...

If truth be told, this whole idea of an empty universe was an embarrassment to most astronomers, who can *see*, every time they look around, that there are mountains of stuff on every side. Nevertheless, the power clique roosting in ivy towers of palaeolithic universities were drunk on mathematical wizardry, and in their metaphysical vision found that "nothing expanding" could indeed be useful to astronomy. Einstein added dust particles to the recipe, and the resulting fusion was the Einstein-de Sitter universe, which, of course, expanded.

This suave, smooth, well-groomed cosmos wouldn't even get onto the radar of consideration for the national rugby team. It is utterly without bumps and scrapes, mountains and valleys, bones and muscle—it is, as they say, isotropic and homogeneous, with zero curvature. That's right, it's flat, and it has no structure. Notwithstanding the foregoing, it is preferred by cosmologists because, at the end of the day, it is the simplest solution to the Friedmann equations, and they obviously cannot do cosmology without solving those equations.*

So, let's answer the question: Is the world outside my window recognisably Einstein-de Sitter? Based on my own measurements, I should think not, but the Standard Model does not question it. Not for a moment. Modern cosmology applies three parameters: The curvature parameter (k); the Hubble parameter (H); and the deceleration parameter (q).

All three *assume* an Einstein-de Sitter expanding model. It was not seen. It was not measured. It was simply assumed.

Whether we like it or not (and there's little doubt that I do not), this is the framework within which we are expected to conduct our measurements. The arbitrary Hubble constant is ever-present as we try to derive distance. During the travel time of light over cosmological distances, the universe would have spread out, and furthermore, in some obscure way allegedly increased the distance involved. (Aside: Depending on who you talk to, expanding space in fact does *not* increase measurable distances, but it seems to me that we cannot logically resolve that conundrum. Sometimes it's better to quietly walk away.)

There are in astronomy a number of techniques in use for estimating distance, reducing naturally in effectiveness as remoteness increases. I shall list them here economically, with a touch of detail and explanation only for those methods applied to extremely remote parts of space. Within the Solar System, we are somewhat spoiled for choice.

Parallax, using the trigonometric properties of triangles, is useful only at very close range because of difficulties in obtaining access to a decently long baseline for the triangle. Just how far

---

* The influence of geometrical curvature on observational astronomy is the subject of chapter eight.

does parallax take us? Limits vary between about 80 and about 300 light years, if we find a 10% uncertainty in the results acceptable. According to a parallax anomaly called the Lutz-Kelker bias, the limit is 50 light years. The European Space Agency launched a bespoke mission in 1989 called Hipparcos to investigate the positions and parallaxes of 100,000 stars. That seems to be the most successful attempt so far, with reasonable confidence up to 300 light years. Although the Hubble Space Telescope could in theory determine trigonometric parallax of stars on the order of several hundred light years distant, in practice the platform is not stable enough for precision astrometry. Notwithstanding a disappointingly limited range, parallax may be regarded as the most reliable method for extra-solar objects too far for radar, even though the attendant uncertainty exceeds 10%.

Radar (bouncing a radio beam off a remote object and timing its return) is excellent, and not only gives us a phenomenally accurate measure of distance and relative velocity, but can be used also to scan surfaces through optically opaque atmospheres and produce relief maps (for example, the surface of Venus).

However, it can be used only on objects that are in cosmological terms right here in the neighbourhood, not least (and not only) because of the turnabout time of a light signal (in terrestrial years, double the one-way distance in light years). A radar map of Andromeda would take about five or six million years to get back to us, and that is the next closest spiral to the Earth's Milky Way.

We can today check the quantities obtained geometrically by using sophisticated satellite laser-ranging equipment, and we are thus able to verify the principles involved. The point I wish to emphasise here is that it was a consequence of comparison between two methods, each proven in practice in our terrestrial environment, which allowed us to adopt these techniques as rungs on the ladder, and equally importantly, permitted us to calibrate them for aberrations and systematic anomalies.

Clearly, direct verification of distance techniques is frustratingly restricted to our immediate astrophysical neighbourhood. *This is the full extent, the very limit, of physically tested methods.* From here on out, we would rely on untested theoretical models to establish the remoteness of celestial objects, and that, unfortunately for empirical science, is where we would expect to find the Hubble redshifts.

The next rung on the distance ladder reverts to the oldest assumption about a source of light—that remoteness is an inverse function of brightness. It is a naïve principle, suggesting that bright stars are closer than dull stars as a rule. There are several problems to be overcome.

We should devote some effort to getting a clear picture of this rung on the ladder, for it is fraught with uncertainties, and that necessarily casts a shadow of doubt onto those rungs that follow, including, ultimately, the use of galaxy redshifts. The period-luminosity relationship of Cepheid variable stars is based primarily on their implicit categorisation as standard candles and so Cepheids joined the now controversial type 1A supernovae as beacons of known value in the cosmos.[*]

Variable stars are amongst the most alluring and mysterious objects seen by astronomers. They dim and brighten with clockwork regularity, and explaining that away challenges our understanding of celestial physics. Without getting into theoretical knots over what causes them to flicker, we can take the Newtonian route and see if we can find anything useful in the effects. It turns out we can.

The period-luminosity method was derived from the discovery by pioneering lady astronomer Henrietta Swan Leavitt in the early 1900s that there is a measurable correlation between the time taken for a variable star to fluctuate and how bright it appears to be—in short, the longer the time lapse between fluctuations, the brighter the Cepheid is to the eye. In principle, it involves determining the distance to standard candles by how strongly they shine.

If it were only possible to determine the distance independently to at least one Cepheid, the intrinsic brightness of the star could be determined. The difference between intrinsic brightness and apparent brightness would then via the supposedly "proven" inverse square law for the fading of light intensity, give us a distance formula for all Cepheids. It was enticingly simple: Get the period and the observed brightness, and you have the distance. This is now an established and much-used astrophysical method, but we should not overlook three important facts:

---

[*] Standard candles are classes of objects for which luminosity data are consistent and known with relative certainty, and are therefore used as marker beacons on the radial axis.

- Although luminosity correlation with rate of fluctuation in Cepheids is clear and easily established, extrapolating that to a reliable distance scale is defeated by the lack of a definite zero point. Because Cepheids lie beyond the range of triangulation, we can at best only infer the distance to any particular example to use as a benchmark.
- Even if we do accept the vagaries of Cepheids as standard candles, they still do not help us with the redshift scale because *Hubble-type expansion theory applies exclusively beyond the scope of period-luminosity tests.*
- The inverse square law is itself of dubious value in this regard. Consider the diffusion of your car's headlight beams on a clear night compared with thick fog. The difference is huge. Cosmologically, we talk of light extinction, and the influencing factors are numerous. We must, in addition to our list of other redshift mechanisms, allow for extinction due to Compton scattering, for example. That becomes impossible to quantify when we realise that most of the Inter Galaxian Medium (IGM) is invisible to current instrument technology.

Much of the groundwork was done by Edwin Hubble. He discovered the first Cepheid variable star, located in neighbouring spiral Andromeda (M31). Using the by then fairly well-established period-brightness rule, he estimated the absolute brightness of M31. Once he had that, the rest was easy—simply apply the inverse square law for the dimming of light to the measured apparent brightness, and distance is given. He got it spectacularly wrong. Hubble's sums indicated that M31 was only 900,000 light years away, whereas today we're pretty sure it's nearly 3 million light years distant.

Quite naturally, on that basis, other magnitude properties for the galaxies in his sample went horribly wrong too. From their angular size and estimated remoteness, Hubble compiled a table of galaxian diameters, starting with M31 and M33. Obviously, there was considerable deviation from true, and his diameters were understated by more than half.

Of course, we have subsequently corrected his data, but new discoveries and advancing instrument technology are more often than not significantly revising previous estimates rather than confirming them. The uncertainties inherent in cosmological distances had started to skew astrophysical data, and the situation deteriorated rather drastically as astronomy reached farther into deep space.

The enforced blackouts of World War II were a delight to astronomers. It was during these periods of near-pristine darkness that Walter Baade found in M31 that there were two distinct classes of stars in the galaxy: Out in the suburban spiral arms where Hubble was originally able to resolve stars, they are bluish, indicating that they are relatively young stars with high metallicity.* These, Baade classified as Population One stars, and our Sun is an example of this class. Nearer the downtown core, the stars take on a reddish hue, have low metallicity, and are old. These are classified as Population Two (occasionally referred to as W Virginis stars), and examples in this category are subdwarfs.

What concerned everyone involved (and hopefully concerns us here) was the fact that Population I Cepheids are *four times more luminous* than Population II Cepheids. The implications were enormous. The extragalactic distance scale was severely shaken by Baade's discovery, and that resulted in a recalibration of the period-luminosity rule originally laid down by Harlow Shapley. It turned out that Shapley had ignored (or been unaware of) absorption by interstellar matter. But that wasn't the crux of the problem. It was the mistaken assumption of Cepheids as standard candles. This would happen again when some famous people tried to convince us that type 1A supernovae are standard candles.

The net result was that Hubble consequently underestimated the luminosity of Cepheids by a factor of four, and that gave him half the correct distance. The bottom line: *The erroneous allocation of standard candles, combined with ignorance of effects in intervening space, resulted in a monumental skewing of the distance ladder for galaxies.*

This leaves us with a dilemma. We actually cannot measure cosmological distance with anything approaching acceptable cer-

---

* The question of stellar metallicity as an aging parameter is puzzling. The general rule is that older stars show lower spectroscopic proportions of heavy elements than young stars. However, by standard theory, fusion steadily converts hydrogen into heavier elements, as far as iron on the periodic table. So the older stars are, the greater should be the abundance of metals, until at the end of their life they would be mostly iron. Therefore, metallicity and age should increase together, not the other way around.

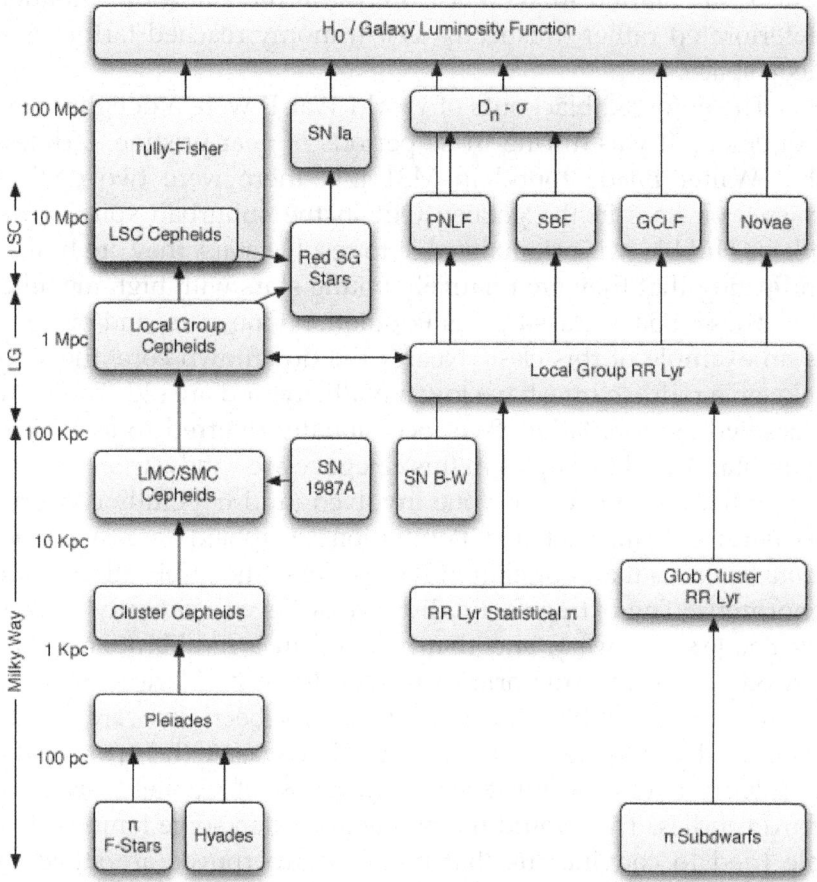

Figure 10: All extragalactic distance techniques are ultimately dependent upon the value of the Hubble constant, which cannot be independently derived from observation. (Originally published in Jacoby et al., 1992. Adapted by Stuart Robbins in 2004 for http://jtgnew.sjrdesign.net).

tainty beyond the reach of parallax (which itself has some daunting attendant uncertainties). That limits us to just 300 light years from Earth. Even that is optimistic—some textbooks set the limit to 80 light years, and taking account of the Lutz-Kelker bias further restricts us to just 50 light years, at 10% uncertainty.

With the period-brightness rule of variable stars we are guessing wildly. Stephen Webb is a professor in the Department of Mathematical Sciences at Loughborough University, and he is recognised globally as a leading authority on the cosmological distance ladder. In his definitive, superbly-written text, Measuring the Universe, Webb has on page 187 a table of distances to Local Group galaxies, based upon Cepheid variables. He prefaces it with the following stern caveat:

> Note also that some of the distances may be inconsistent with other values given in this book. This reflects the fact that distances on this rung of the ladder are uncertain! Different workers obtain different values. [*]

He is not in the least bit exaggerating the case. The Cepheid-based distances are in some cases accepted with 50% uncertainty! Period-brightness takes us out to (very optimistically) 30,000 light years. To put that in perspective, the Sun is somewhere between 20,000 and 30,000 light years from the centre of our home galaxy. So, given the potential vastness of our field of endeavour and the great uncertainties built into it, we should concede that we are really nowhere with the distance ladder.

It was upon this rather dizzying platform of period-brightness that Edwin Hubble found himself when he launched his study of offshore galaxies. He framed his efforts around some questionable assumptions that he would later regret, but we shouldn't hasten to condemn him for this. It reveals a pervasive weakness in the methodology of space science that results directly from our training at university.

This is terribly important. I wish to recall something I said in the first chapter: Newly graduated astronomers experience a rush of blood. In our eagerness to conquer that darkly beckoning, windswept desert of ignorance we call the Universe, we quite forget to interrogate our assumptions. Some of the most lauded scientists of all time have been guilty of this, and Edwin Hubble was no exception.

He first read law as a Rhodes Scholar at Oxford University before returning to the United States. There he had something of an epiphany which led to his abandoning law for a doctorate in astronomy at Yerkes Observatory, home to the world's largest refracting telescope. As a postdoc, he moved to the Mount Wilson Observatory and a five year stint on the hundred-inch, the world's largest *reflecting* telescope. In 1924, just 5 years after he was capped, he published his first results on offshore galaxies. It

---

[*] Stephen Webb, *Measuring the Universe — the Cosmological Distance Ladder* (Chichester: Praxis Publishing, 1999).

was clear that Edwin Hubble was riding on the flush of a pink cloud. He knew where he wanted to go with all this.

At this point, we can rely once again on Stephen Webb to give us an honest appraisal of the status quo regarding galaxian distances. Before we take the next bold step and cast off into the uncharted sea of cosmological redshift, it would be wise to heed his warning:

> These galaxies are key objects in calibrating the rest of the extragalactic distance ladder. The disappointing conclusion is that, although some astronomers would argue otherwise, these distances are probably uncertain at the 10—20% level...Now that we have reached the Mpc scale the distance ladder starts to look shaky.

Indeed it does, but there's much worse to come, I'm afraid. Webb goes on,

> Each step on the distance ladder introduces further uncertainty. Would it not be better to use primary indicators to calculate galaxy distances, and thus remove the need for the treacherous distance ladder?

These words clearly highlight the problem—we simply don't have primary distance indicators applicable beyond, at most, a tenth of a kiloparsec, let alone to the scale of Megaparsecs. The core issue is that there is no overlap. Parallax falls just tantalisingly short of Cepheids. How different things would be if we could only anchor each succeeding distance formula upon the certainty of an empirically solid foundation in the one prior to it. But we cannot, not yet anyway, so we have to make the best of what we've got. [*]

A word about masers: The discovery about 20 years ago of water-driven masers near the galactic core heralded the onset of some optimism amongst astrophysicists. Masers are transient phenomena, coming and going regularly around the centre of the galaxy. By employing extra-long baseline radio interferometry, we may be able to get reliable measures over tens of thousands of light years. [†]

---

[*] Reminder: A parsec is an astronomical unit of measure equal to 3.26 light years. The prefixes *kilo* and *Mega* mean *thousand* and *million*, respectively. When in doubt, please refer to the glossary and addenda.

[†] MASER is an acronym for Microwave Amplification by Stimulated Emission of Radiation. It is a type of laser, with similar properties.

In 1988, Mark Reid and colleagues used the proper motion of a maser in Sagittarius (near the hub of the Milky Way) to infer a distance from the Sun of 7.1 kpc, about 23,000 light years. Although the technique is commonly labelled a *direct* measure of distance, it is nevertheless inferred from quantum theory, and gives an uncertainty of more than 20%. In any event, masers have a long way to go yet before they can conquer Megaparsec scales.

Similar quantum techniques were standard practice for decades in estimating the disk diameter of stars, which until a couple of years ago, had been optically detectable only as mere points of light, even through the most powerful telescopes on Earth. Recently, the optical disk of the Red Giant star *Betelgeuse* was resolved for the first time, and we now have the means to check our prior methodology. I am keenly awaiting results from the first person brave enough to do so, but no such luck yet. I may just have to do it myself.

Recently, Canadian mathematician Bruce Rout made a sterling effort to devise an independent measure of galaxian distances using the underlying General Relativity Theory (GRT). What he did was extract from Einstein's gravitational theory a relationship between spiral morphology in galaxies and their rotational velocity. At the suggestion of his wife Roxanne he crunched the numbers and gave a formula suggesting remoteness in space from the angular separation of spiral arms. Hence, we have both the "Rout Spiral Relation" and "Roxy's Ruler". These results were published in three papers, The Spiral Structure of NGC 3198 (*viXra*: 0909.0007), A Comparison of Distance Measurements to NGC 4258 (*viXra*: 0911.0016), and Distance, Rotational Velocities, Red Shift, Mass, Length, and Angular Momentum of 111 Spiral galaxies in the Southern Hemisphere (*viXra*: 0911.0023). All make for fascinating reading, but I shall concentrate on the last-mentioned here. His conclusions are dismal for the Hubble Law and universal expansion models, I'm afraid.

> This paper makes direct measurements of distant galaxies by comparing spiral arm structures to the expected locus of gravitational influence along the geodesic in a centripetally accelerating reference frame. Such measurements provide a method of independent validation of the extragalactic distance ladder without presupposition of the uniformly expanding universe theory. The methodology of this paper avoids the use of Hubble's constant in the measurement of

the distance to galaxies beyond the range of contemporary direct measurement methods... A Hubble diagram calculated using this method is presented from data obtained from 111 spiral galaxies in the southern hemisphere to about 200 Mpc distance. The galactic red shift from these galaxies appears independent to distance.

The repertoire contains a great many less-used techniques, which either specifically or by adaption contribute to the spatial map of the known universe. I have selected those I consider to be the most important for this review. Although they do form part of the distance ladder, gravitational lensing, the Sunyaev-Zeldovitch effect, and the extremely robust Tully-Fisher Relation will be discussed within the context of ensuing chapters.

That brings us, at long last, to the crucial rung on our ladder called "cosmological redshift". Thank you for being so patient. It'll have been worth the perseverance, I'm sure, when we see egg on some bashful faces. It all started off rather well, and cosmologists at the time had every reason to be cheerfully optimistic. In 1912, an astronomer named Vesto Slipher at the Lowell Observatory found that the spectrum of Andromeda (M31) was noticeably blueshifted.

This intrigued him, for it indicated (by applying the Doppler principle) that M31 was rushing towards us. The prospect of looming catastrophe motivated him to examine the spectra of 14 more galaxies, and it transpired that of the 15, thankfully, only two were blueshifted. The rest were redshifted. They seemed to be going the other way.

Now, this should not surprise us in the least. Of course we should expect to see vastly more cosmological redshifts than blueshifts. It's obvious; it is far easier, with many more opportunities at every turn, to *lose* energy than to *gain* energy. Think about it. What happens more spontaneously, with less effort—water coming to the boil or water cooling to room temperature?

In due course, Slipher's database firmly caught the attention of one Edwin Powell Hubble at the Mount Wilson Observatory. Assisted by Milton Humason, Hubble diligently sought out the spectra of 18 offshore galaxies. To their great satisfaction, the plot of redshift against distance was a straight line, and became, in the words of Stephen Webb, "thereby one of the greatest discoveries of the century." This confirmed the prior conclusions of Carl Wirtz and Knut Lundmark. American physicist Howard Robertson became the first to use these data to formalise the relationship between velocity (that is, Doppler redshift) and remoteness.

It seemed that a controversial cosmology that had travelled a tortuous route from Gauss to Lorentz and Poincaré, from Albert Einstein to Alexander Friedmann and thence to Georges Lemaître (all without a single observation between them to cloud their judgement) was vindicated. The Universe was indeed expanding, and we could now map the process with galaxian redshifts.

It seemed so cut-and-dried. The idea made such marvellously good sense; it just *had* to be right! Not so?

Well, actually, *no*, absolutely not. The principles of good science had been beaten yet again by unseemly haste, this time with absolutely awful consequences for astronomy. In amongst words of glowing enthusiasm for Hubble's messianic revelation, Webb tucks just the tiniest little bit of caution:

> And as we shall soon see, there were theoretical reasons that there should be a velocity-distance relation. Like so many other discoveries in science, the idea was 'in the air'. Hubble receives the credit, though, because it was his weight of evidence that put the matter beyond doubt.[*]

Ah, what delicious irony! Beyond doubt? Would that there were such things in science, Dr Webb. The illusion of infallibility that consorts with elegant mathematics had been intoxicating, and everyone was drunk. You see, it turns out that there *was* no relationship, neither theoretical nor observed, between redshift and velocity in the sample of galaxies under review. It was an illusion. Edwin Hubble and the excited crowd around him had looked at the data through the rose-tinted spectacles of assumption. They *assumed* the model *before* they worked the numbers. Fatal error!

All of the galaxies in his studies were local, and nearly all fell within the ambit of a cluster known as the Local Group. Hoyle, Burbidge, and Narlikar, in A Different Approach to Cosmology, pages 32—33, made the decisive observation that,

> In the case of the redshifts it had been accepted that they must be corrected for solar motion with respect to the centroid of the Local Group, since it had been realised since 1936 that the systematic redshift does not operate within the Local Group.

---

[*] Stephen Webb, *ibid*, p. 241.

In my opinion, the last few words in the quotation above are amongst the most important statements ever made in criticism of modern cosmology: *Systematic redshift does not operate within the Local Group.* The essential point is that although no real relationship existed, it was nevertheless "found" in the data. Let us consider this very carefully.

The assumption, quite arbitrary and obvious in hindsight, skewed Hubble's judgement. Winner of the 1979 Nobel Prize for Physics Steven Weinberg looked at Hubble's sample, and had this to say:

> In fact, we would not expect any neat relation of proportionality between velocity and distance for these 18 galaxies—they are all much too close, none being further than the Virgo cluster. It is difficult to avoid the conclusion that...Hubble knew the answer he wanted to get.[*]

In any event, it is absolutely astonishing to consider that an assumption—that there is a relationship between redshift and distance—was made with such impenetrable certainty that an entire theory of the Universe could be based upon it, yet it was within a few years clear to all, including even Edwin Hubble himself, *that such a relationship did not exist in any shape or form within the data set from which it was originally drawn.*

Because expansion is said to take place only far away, we have no means of relating distance to redshift. Close by, in the Local Group, where we have at least some means of guessing at distance, there is absolutely no correlation with redshift. In fact we have blueshifted galaxies, so if redshift-distance applies, when we look at Andromeda, it is actually behind our heads! No better example of explicit, model-driven bias exists anywhere in the long, sad story of modern cosmology.

In 1983 Geoffrey Burbidge, who was at the time Director of the Kitt Peak National Observatory, commented as follows:

> There is at least one well known physicist, in fact Dr. Varshni, who remains unconvinced that the redshifts are really present, and interprets these lines as various elements at rest. This changes the whole picture and Dr. Varshni is at one end of a spectrum of argument associated with these questions. If there are no redshifts, this is another way of saying that we don't really know much about the Universe outside our own galaxy and the immediate

---

[*] Steven Weinberg, *The First Three Minutes* (New York: Basic Books, 1977).

vicinity. So, it is a somewhat radical approach and you really have to go back 50 or 60 years (1920s) and start re-thinking a lot of questions ... I must tell you that extraga-lactic astronomy is really in the form of an inverted pyra-mid, with a small number of facts at the base on which a large superstructure has been created, some of this super-structure may indeed fall down.

Despite the complete invalidation of the original galaxy-redshift data set, it was never replaced. No further "curve fitting" was achieved, and I doubt whether it has even been attempted. In my view, it is in any case not possible in practice, because, as we have seen already and shall confirm in the pages ahead, the mooted expansion takes place exclusively in deep, deep space, far beyond our ability to empirically verify it. Please remember the following whenever you are confronted with confident assertions by cosmologists that the Hubble redshift relationship is simply curve fitting, the automatic result of making a plot with meas-urements. Given that uncertainty increases dramatically with re-moteness on all axes, it would appear that *the Hubble relationship fits best where it is tested least.*

I have emphasised this because it is the mantra that demysti-fies modern cosmology. It is arguably the most important sen-tence in this entire book, pregnant as it might be with important sentences. The Hubble law, as controversial in its assertions as the choice of Hubble's name to label it was unfortunate, is laminated upon the intricate and convoluted surface of the consensus model. Please help me to peel it away.

Discussion: At the scale upon which Hubble-type expansion is alleged, distance measurement using redshift is unverified and totally unreliable. Cosmological redshift is not a calibration of distance, and does not lend support to or justify expansion the-ory. The Hubble law is a myth.

Chapter 4

Redshift

Put it where it fits best

The Hubble law is a systematic displacement in spectral lines taken to mean recessional velocity. "Needless to say these values (quasar velocities) are without physical significance and clearly indicate that the cosmological red shift hypothesis is completely untenable."

(Dr Y.P. Varshni, University of Ottawa).

◆ ◆ ◆ ◆ ◆ ◆ ❖ ◆ ◆ ◆ ◆ ◆ ◆

The suspicion that light energy degrades to lower frequency as it travels was given impetus by an unexpected discovery on the Sun. It's called the "solar centre-to-rim redshift". Doppler shifts in sunlight are first corrected for rotation (redshift on the receding limb and blueshift on the approach). The residual redshift displays an interesting anomaly — *the redshift of light from the centre of the Sun is less than that from the limb.*

This astonishing fact is key to understanding light and how redshift is manifested. The only difference in initial properties is that light from the middle of the Sun covers a smaller distance getting to us than light from the extremities, by an amount equal to the curvature of the Sun's surface. Relativistic gravitational redshift is independent of centre or limb, so the only conclusion solar physicists could reach is that light travelling the greater distance passes through more space, and thus interacts with more

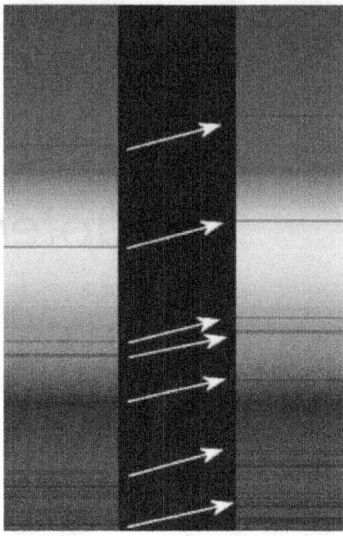

Figure 11: The term "redshift" describes a deflection towards the red end of the spectrum (at the top of this diagram). It assumes sunlight as the benchmark. (Wikimedia Commons public domain image by Harold T Stokes, amended by Ian Tresman).

particles. It is a clear observational demonstration that light loses energy as a function of travel time through a particulate medium.

Therefore, in this sense, all cosmological redshift is tired light. As time goes by on these fabulous journeys across the Universe, light dims down, the waves flatten, and it transports less and less energy. Since energy is conserved, whatever bleeds off in interactions along the way has to go somewhere, and it has to do it without necessarily blurring the images.

Light gets tired, and its weariness is measured by a quantity called redshift.

There is a caveat: An indication in the ejection of quasars from active galaxies suggests that the reverse may be true. It's possible that light *increases* in energy over time, in other words, that it *blueshifts* off a higher base than sunlight. We shall consider this possibility in the next chapter. For now, let's concentrate on recessional motion. The notion of universal expansion is the bedrock of modern cosmology. It is calibrated by the implied Doppler shift in celestial spectra. I don't buy it. There is some evidence—not much, admittedly, but some nevertheless—indicating that the signature spectrum of a radiant object changes with age, or from the rigours of travel in a hostile world, or both. The example of solar redshift cited above illustrates the point. There are many more examples, from observation, experiment, and theory.

We are perfectly within our rights to be curious about cosmological redshift, but let's face it; we can't take it very far. The

Achilles' heel of redshift astrophysics is lethal: We have no means at all of measuring the degree of change in the positions of spectral lines, for the simple reason that we are unable to establish just where they were originally. We are damned by the very remoteness we seek to master.

A thesis, such as I hope this to be, that presents the case for a non-expanding Universe within the context of modern astrophysics is really just opposition to the Standard Model of Cosmology. Furthermore, contesting the expansion hypothesis rests almost entirely upon critical examination of cosmological redshift (given that for now, we ignore the role of geometrical abstractions). Let's deal with the basics first, and if we can, save geometrical arguments for chapter eight.

We commonly assume that cosmological redshift is entirely a Doppler effect; that is, a stretching of the light wave caused by the source moving away from us. Ironically, this is one thing that both sides to the Big Bang debate eventually agreed upon: It is *not* a Doppler shift. The conventional explanation was stumped by the inevitable theoretical catastrophe contingent upon cumulative recessional velocity—that it would soon exceed the speed of light. That's taboo! They had to think fast. They substituted a relativistic, de Sitter-type, expanding, non-metric space, and breathed again. Their theory lived to fight another day. Let's face it; it's tough at the top.

The pivotal issue in all matters pertaining to our current understanding of the transmission of light, from Maxwell's equations of electromagnetism through to the Special Theory of Relativity, is that of *vacuum*. The genius of James Clerk Maxwell lay in his solution to the seemingly insurmountable problem of wave propagation in empty space. He established a principle upon which SRT could proceed. All effort was aimed at dealing with the thorny issue of empty space, and in a sense, it became self-congratulatory: Maxwell and Einstein, Poincaré, Hilbert, and de Sitter, a veritable A-team of mathematical theorists swept the problem aside. In a somewhat crass denigration of empiricism, none of them bothered to experimentally verify the vacuum that is assumed.

Here is one of those defining moments in the roll-out of ideas in this work. Once again, it seems, enormously talented scientists applied their brilliance to a persistent obstacle in their path, and jointly defeated a problem that in reality, simply doesn't exist.

Question: Where is the much-vaunted vacuum? Answer: Nowhere in sight. The vault enshrouding our spaceship Earth is teeming with life; there is dust, gas, radiation, plasma, neutrinos, electricity, and magnetism—and gravitation, of course—subtly or grossly violating the pristine emptiness of every nook and cranny.

Andre Assis, in his excellent work "On Hubble's Law, Olbers' Paradox, and the Cosmic Background Radiation" cites the reasoning of radio astronomy pioneer Grote Reber:

> On the other hand, as has been pointed out clearly by G Reber (1986), the main reason for adopting the hypothesis of a Doppler effect as the cause of redshifts has been the assumption that intergalactic space is a void and that nothing happens to light in its journey from a galaxy to the Earth.[*]

There isn't any patch of space, anywhere, in any direction, that does not have measurable energy-density. Nowhere, whether by observation with astronomical instruments, or by experiment in the laboratory, do we find a vacuum. Absolute zero on any axis or scale is a theoretical dream, and though we may approach it with tantalising proximity, it always eludes us in practice. "Nothing" is just not physically possible.

We may profit by considering space as generally gaseous. There is for the most part an ultra-low-density cloudy distribution of energy and particulate matter, interspersed with vastly separated, high-density clumps. In our field of investigation, the clumps would typically include nebulae, stars, galaxies, and clusters. As far as the propagation of waves is concerned, we can clearly follow the passage of unfolding shockwaves in interstellar space, even in areas that appear to be void of a medium capable of sustaining pressure waves. We know that light speed and energy vary by the media in which it travels, and the Earth's atmosphere gives us a wonderfully accessible laboratory with which to quantify the effects.

We have already discussed this in sufficient detail, so let's leave it there for now. We all agree. Cosmological redshift is not a Doppler effect. It's a universal Doppler illusion. Well, no theory is perfect. We all have our problems. The elders of cosmology might conceivably have got away with it, if only they had left space as it

---

[*] A. K. T. Assis, "On Hubble's Law, Olbers' paradox, and the Cosmic Background Radiation" *Apeiron*, No. **12**, Winter 1992.

is, and postulated a normal explosion to push the galaxies apart at a decreasing rate in a finite universe. They would in that way, I think, have more reasonably argued their case. Then solve the problem of infinity by suggesting cycles, and you've got a half-way decent cosmology. Now, isn't that much more sensible?

The term cosmological redshift refers to a measurable shift in spectral lines of radiant objects at cosmological (that is, very great) distances. Redshift is a trend towards longer wavelength in light, marked by a shift towards the red (long wavelength) end of the spectrum. It indicates a lower level of energy in the light wave when compared with a laboratory standard, generally taken to be the spectrum of sunlight as measured on Earth.

This shift is seen and measured by astronomers, so we are not contesting the existence of redshift. That would be foolish. What we are actually having trouble with is twofold: Shifted in respect of *what* exactly; and the currently dominant *interpretation* of the effect. It is really a matter of choice. There are five suggested causes of frequency change in light waves:

1. The Doppler Effect, the result of approach or recession.
2. Scattering and friction—physical interaction between light and obstacles in space.
3. Gravitation, usually but not in all cases, curvature of spacetime *a la* General Relativity.
4. The expansion (creation) of space itself, which broadens the light curve.
5. Ageing.

All five could jointly or severally cause redshift, although points 4 and 5 are controversial, and avoided by some scientists. In light that has travelled a long time from very distant objects—it can in principle amount to an elapsed time of many millions of years—the redshift should be a compound effect added to whatever the spectrum looked like at the outset. We need first to clearly grasp that frequency deflections from laboratory standards are really the flavours in a well-mixed cocktail. Our job as astrophysicists is to unravel the blend and write down the recipe. If we are obedient to the constraints of proven physics, we must concede that redshift in the spectrum of a distant object really *is* a cocktail, not a single ingredient. There may well be a Doppler component in the overall mix, but in what proportion? We cannot say.

We can see the effects, but we are groping wildly in the dark trying to establish the causal mechanisms. Before we start to calibrate cosmic redshift, however, it would befit our objectivity to determine whether or not we have a standard arrangement of spectral lines for young light. Without that benchmark, it is impossible to calculate how much shift has taken place, and of course, the whole Hubble relationship becomes science fiction.

Alexander Boksenberg is an eminent English astronomer and physicist who invented a photon-counting device that significantly improves the optical power of telescopes, and he used this to study objects with very high redshifts. He carefully examined the absorption lines in spectra from quasars, and discovered something of great significance to cosmology: The lines did not relate to the source of the light (the quasars themselves), but to the galaxies and inter-stellar dust that it passed on its journey to us. [*]

Therefore, redshift in the spectrum of a quasar is not source data, but rather a mere factor in a set that includes whatever else interacts with the light over time. This effect makes getting the real picture from spectra so much more difficult, and introduces a plethora of variables to be considered before one can reach a conclusion. Unfortunately, there is a consequent subjective influence that has crept into spectroscopy, and it has moved significantly from science to art. Most redshift measurements simply fail to properly take this complexity into account, and as a result many catalogues are completely skewed.

A greater redshift may only coincidentally and tenuously imply greater distance. The seeming tendency of light to redshift as it ages is one of its most confounding properties. When light leaves a source, pristine and untainted, it is presumed to have a spectrum identical to one we produce from a local light source in the laboratory. All the spectral lines and colours are taken to be in their benchmark positions, crouched in the starting blocks as it were.

In other words, the frequencies (and therefore wavelengths) of the colours are within the parameters of what we consider the native state of light. From that moment on, things change, but because we are working with such microscopic measurements,

---

[*] Of course, Boksenberg's device is also used in the Hubble-Tolman number test (see chapter two) to observationally verify a non-expanding Universe.

these changes become significant only over astronomical distances. It is quite impossible to test this hypothesis in the laboratory.

There is a word that has been raised to alarming popularity in physical science, especially anything to do with the much-maligned heavens above. It is the noun *anomaly*, often used as the adjective *anomalous*. It indicates something that deviates from a standard or norm, and more particularly infers an unexpected result. The word "anomalous" is so prevalent in cosmological redshift studies that one is brought to wonder how science could be conducted this way.

Here, "anomalous" is first and foremost a synonym for "intrinsic". Intrinsic redshift (that is, a redshift at source that is higher than our sunlight benchmark) is strongly indicated by observation—much more strongly, it need be said, than the Hubble effect—and theorists are scratching their heads to find the physical cause of the phenomenon. To their eternal credit, they have disdained the now acceptable explanatory technique of assigning the cause of delinquent situations to supernatural, infinitely tuneable, "dark" stuff.

Without going into conjecture about how nascent systems can acquire embedded redshift in the first place, let us review the evidence. To assess whether the arrangement in an apparent system is or is not anomalous, we would, in the words of Arp and the Burbidges, look for "properties of nearness, alignment, disturbances, connections". Thus, we may assume that there is something anomalous about measured redshift if:

- There is a prevalence of high redshift objects near the nucleus of nearby galaxies, or high redshift galaxy-like systems associated with low redshift clusters;
- Physical connections are seen between objects with significantly varying redshifts;
- Apparent proximity of high redshift objects is given by non-redshift distance indicators;
- Radial alignment suggests ejection and common origin of objects with excessively varying redshifts;
- Higher redshift objects appear in the foreground of lower redshift background systems;
- Several other more technical clues, like morphological associations and the Karlsson Effect, which for reasons of

brevity we shall deal with only superficially in the present investigation.[*]

The entire gamut of anomaly indicators listed above is seen abundantly in observation, and although publication of anomalous results is covertly discouraged by mainstream journals, there is enough published evidence to prove the point. My own summary paper presented at the national Symposium of the Astronomical Society of Southern Africa in Durban, 2008, entitled "A Review of Anomalous Redshift Data", carries a schedule of nearly 60 useful references from the literature.[†] The peculiar properties of quasars, and their role in presenting intrinsic redshift, are dealt with in the following chapter, so I shall exclude them from the current discussion. I think the principle is established well enough in the foregoing pages without now having to list examples; the journals of space science are awash with them. With the exception of papers by known blackguards (like me, I suppose), most are available on *arXiv*.

Our investigation in this chapter is concerned with any real or perceived pattern in the magnitudes of celestial redshifts. I'd like to get the Karlsson Effect out of the way first, since it is, despite being tremendously important, quite independent of the Hubble law, and the Hubble law is what has everyone in a tizz about expansion. If the energy levels of cosmological light are really just a function of remoteness, given that the Big Bang model postulates a smooth distribution of matter in the expanding universe, then we would expect that redshift values should present without digital breaks. The tabulated values would appear randomly, reflecting the suggested patternless distribution of light sources in the cosmos. If on the other hand, redshift relates somehow to the internal energy of the source, perhaps even to mass fractionation layers within incandescent objects, then we might expect something entirely different. Speculation aside, the Lambda-Cold Dark Matter model does not accommodate periodic redshifts. Are they observed?

---

[*] The Karlsson formula is $(1 + z_2)/(1 + z_1) = 1.23$, where z_1 and z_2 are consecutive redshift values.

[†] Hilton Ratcliffe, "A Review of Anomalous Redshift Data" *vixra*: 0907003. It will appear in Franklin Potter (editor) *Proceedings of the 2nd Crisis in Cosmology Conference*, ASP Conference Proceedings (2009), and in the *Journal of Cosmology*.

NGC 5985

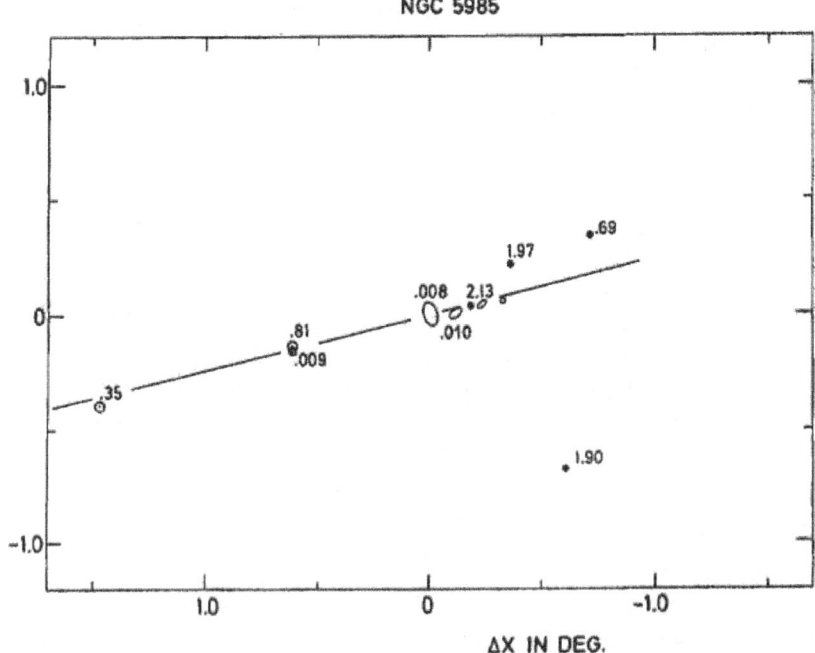

Figure 12: Double whammy! Visual evidence of 5 quasars linked to a Seyfert galaxy, all with quantised redshifts according to the Karlsson formula (Diagram by kind permission of Halton Arp).

The Karlsson Effect refers to certain values in the redshifts of cosmological objects that appear more commonly than others. Preferred values in quasar redshifts were first detected by Margaret and Geoffrey Burbidge in 1967. Four years later, K.G. Karlsson confirmed the effect and derived a formula that constrained the periodicity. That earned him the honour of having his name pinned to an observed effect that was, in the words of Chip Arp, "one of the truly great discoveries in cosmic physics."

Three decades later, with far more comprehensive catalogues of data to work from, Doctors Burbidge and Napier published a summary of the by now overwhelming evidence for redshift periodicity entitled "The Distribution of Redshifts in New Samples of Quasi-Stellar Objects." The authors declare:

> Early in the studies of QSOs, a sharp peak in the redshifts at $z = 1.955$ was reported (Burbidge & Burbidge 1967). Soon after this it was claimed that if we restrict ourselves to low redshift QSOs and related objects with similar optical spectra, now called AGN, the redshifts show a quantized ap-

pearance at values of $z_n = n \times 0.061$, at least up to $n \simeq 10$ (Burbidge, 1968). Initially, with only 70 objects known, a strong peak was seen at $z = 0.061$, and this has persisted with more than 700 objects measured with $z \leq 0.2$ (Burbidge & Hewitt 1990) [...] The results obtained in this paper together with the earlier work, and the statistical evidence for associations between galaxies with comparatively small redshifts and QSOs with large redshifts suggest that QSOs with intrinsic components are ejected from galaxies. If the sample is comparatively nearby so that z_c is very small the intrinsic redshift z_i will dominate and this periodic effect will be seen. For QSOs ejected from galaxies with non-negligible values of z_c the periodicity will not be seen because of the smearing due to the cosmological term. However for properly chosen samples of QSOs and galaxies the clustering tendencies will still be detectable.[*]

Of course, there was considerable resistance to the discovery of periodic redshifts, and the discomfort is understandable. The upholders of orthodox law in cosmology quite naturally hoped that the effect would be local only, and they might then explain the patterns by "cosmic structure in the direction of the north galactic pole" which "probably would have caused such effects" (these are the words of an anonymous referee rejecting the submission of my paper on anomalous redshifts). No such luck, I'm afraid. Wide area surveys like the Sloan Digital Sky Survey (SDSS) soon presented ample data confirming the Karlsson effect for as far as we are able to measure.

In 2009, Martín López-Corredoira presented a summary of quasar anomalies in his paper Pending problems in QSOs.[†] He clarifies the point:

> However, Hawkins et al. (2002), Tang & Zhang (2005, 2008) found that there is no periodicity of QSOs in SDSS and 2dF surveys beyond randomness and selection effects. Napier & Burbidge (2003) argued that Hawkins et al. had not measured the redshifts of these faint quasars with respect to the redshift of their active parent galaxies. The periodicity found of the redshifts is measured with respect to the parent galaxy; people who do not find a periodicity would simply measure the redshift with respect to us ($z = 0$).

---

[*] G. Burbidge and W.M. Napier, "The Distribution of Redshifts in New Samples of Quasi-Stellar Objects" *arXiv*: astro-ph/0008026.

[†] Martín López-Corredoira, "Pending problems in QSOs" *arXiv*: astro-ph/0910.4297).

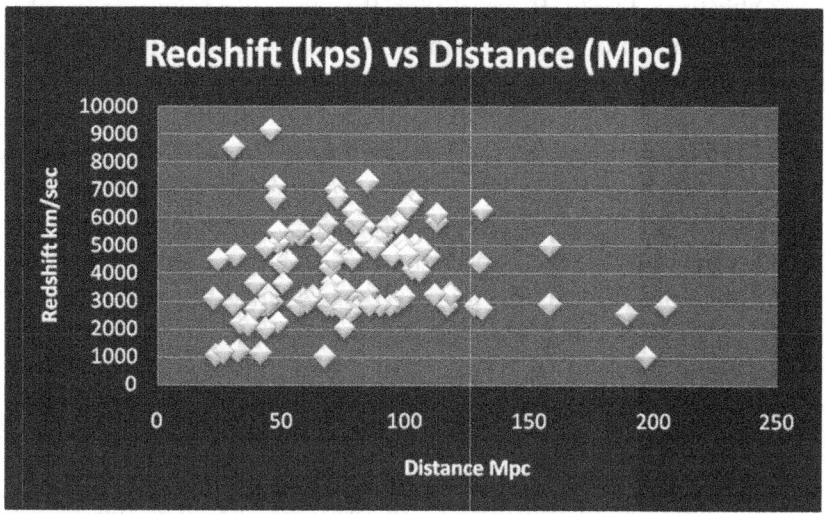

Figure 13: A plot of redshift versus distance for 111 galaxies derived by means of Rout's Spiral Relation. Several frontline astrophysical journals refused to publish Rout's results, leading to speculation that anomalous results are being suppressed.

In other words, the periodicities do exist in the SDSS data if the base value taken is the host galaxy's redshift, and not $z = 0$ as used by the studies that found no unusual preferred values.

Here is an example. In a recently published study, John Hartnett of the University of Western Australia found a distinct periodicity in the redshifts of quasars in the 6th annual release of data from the Sloan Digital Sky Survey (SDSS). Hartnett succinctly summarises his results thus:

> Recently Bell and McDiarmid (2006) analyzed the data from the third data release of the Sloan Digital Sky Survey (SDSS) and found a significant peak in the power spectrum near $\Delta z = 0.62$. Here Δz represents the periodic interval seen in quasar redshift abundances. In this paper I analyze the SDSS sixth quasar data release using a Fourier transform of their redshift abundances as a function of redshift. I show, regardless of any interpretation of the meaning of the redshifts, and aside from any cosmological assumptions, that there is a significant periodicity in the SDSS quasar redshift abundance data.[*]

[*] J. G. Hartnett, "Unknown selection effect simulates redshift periodicity in quasar number counts from Sloan Digital Sky Survey" *Astrophys. Space Sci.* **324** (2009), pp. 13–16; *arxiv*: astro-ph/0712.3833.

Although Hartnett concludes that the periodicity is a selection effect and not some intrinsic property of the quasars themselves, it is nevertheless startling that quantisation of redshifts is so apparent in the data, yet pointedly ignored in the Standard Model of Cosmology. In fact, in 2009 two subsequent papers by Hartnett on the same subject were imperiously rejected by Cornell University's supposedly neutral *arXiv* online archive database, even though they had been peer-reviewed and accepted for publication in journals. In just two years since the paper quoted above appeared on *arXiv*, the crackdown on observations that might cause discomfort for the expanding universe model has become stiflingly more stringent. Welcome to the Brave New World!

I'd like to suggest that Paul Ginsparg's team at Cornell are in their blatant censorship inadvertently doing us a great favour. If anyone engaged in the pursuit of pure science ever has doubts that political suppression of non-aligned results really does exist, all they need to do is take a look at how the *arXiv* administration treats individuals and papers that do not conform to their preferred ideology. When the paradigm eventually falls, the anonymous *arXiv* moderators will, to their horror no doubt, have played their part in its destruction by the very fact that they were unashamedly the vanguard of thought police defending the bastion of consensus cosmology. History has shown that the activities of those who use their power to entrench a ruling clique at the expense of vulnerable individuals who might disagree with them are not immune from public scrutiny, and are seldom remembered with fondness by generations that follow their demise. It will come back to haunt them.

Evidence mounted up in measurement after measurement, yet denial on the part of entrenched astronomers was resolute and inflexible. Martin Rees, who at that time directed the Institute of Theoretical Astronomy at Cambridge University, led the charge. His position was understandable. After all, his entire career in space science had been built upon the notion of universal expansion. He has written best-selling books on the subject. It would be totally unreasonable to expect him now to recant and turn in both his doctorate and his knighthood just because his ideas were in conflict with observation. Notwithstanding his valiant efforts to contain the revolt, it soon became obvious beyond reasonable doubt that cosmological redshifts are quantised. To deny it would be akin to contesting that the sky is blue. Yet it is denied with venom. How sad for science...

There are a number of ideas in the hallways of science regarding the ageing of light, and why, in an apparent contradiction of Newton's laws, a wave motion in space should tire and lose frequency. The Compton Effect, named after its author, American physicist Arthur Holley Compton, gives the clearest answer. It refers to an amazing property of colliding particles and the transfer of energy when collisions occur.

The Inter Galactic Medium (IGM) is a sea of radiation, interstellar plasma, and sundry particles like neutrinos and electrons. Photons, in common with material particles, have energy and momentum, and additionally boast wave characteristics like frequency and wavelength. The higher the frequency of a photon (that is, the shorter its wavelength), the greater is its energy. The critical questions are, should we expect that photons will interact with other particles as they travel through interstellar space, and if so, will there be an exchange of energies?

The literature is replete with plausible explanations for tired light. Not one of them, as far as I know, is without its critics, and I have personally attended vigorous debates on this issue at the *Crisis in Cosmology* conferences. If only one iota of consensus has been reached about tired light, it is this: It is explained in every case I am aware of, with sound physical science and logic. If the argument falls foul of the laws of physics and chemistry, it is rejected. No latitude is allowed in terms of conservation of energy and thermodynamics. Already, in my judgement, that puts tired light way ahead of the standard explanation of cosmological redshift. There is simply no race when it comes to comparing the standards of physics employed; tired light wins hands down.

To my knowledge, the first really promising explanation employing orthodox physics came from the late Professor Paul Marmet. Studies of optical properties of images that have passed through the Earth's atmosphere revealed the principle: If practically all photons pass straight through the atomic structure of a gas without scattering away from their original trajectory, how *do* they lose energy? The answer is quite technical but well understood and clearly seen in the laboratory. When electromagnetic radiation passes over an atom, the atom becomes transversely polarised with nucleus and electrons being attracted in opposite directions. Part of the EMR energy is passed along the atomic axis, stimulating electrons to emit extremely low frequency photons perpendicularly to the light path.

This secondary light stream resulting from the slowing down of light by atoms is called *bremsstrahlung*, German for "braking radiation." The carrying away by transverse radiation of light energy lost in collisions with atoms would cause a redshift along the original light path without blurring the image. The only caveat is whether intergalactic space has a high enough density of atoms; we just don't know.

The "Gravitational Viscosity" hypothesis put forward in 1992 by Ernst Fischer is intriguing. Moreover, it is plausible, uses good physics, works in a non-expanding frame, does not blur images, and would coexist happily with any other proposed energy-loss mechanism to make up the total redshift presented by a cosmological object. Nothing much seems to have happened with this idea since then, and that's a pity. It's worth considering.

In the abstract of his paper he states,

> Applying the basic concepts of general relativity to the global motion of a particle in a mass-filled universe leads to a loss of momentum relative to the rest frame of the universe. This loss is caused by the different running times of the gravitational interaction quanta exchanged with masses in front and behind the moving particle, if the signal velocity is limited to the speed of light. Due to this 'gravitational viscosity' of space, the energy of photons will be reduced with the time, and thus with the distance of the emitting source.[*]

In a Universe where Mach's principle applies (that is to say, space infused with mass, and everything connected gravitationally across the whole cosmos), the photon would be fighting its way constantly through gravitational resistance. It is really a generalised Sachs-Wolfe Effect, but without invoking expansion. It is a purely linear effect caused by gravitational differential fore-and-aft.

The idea is delightfully simple, and I'm hoping it will reignite interest in the concept of drag scrubbing off energy in space. It hints at the aether, another idea whose time may finally have come. There does seem to be some revitalised interest, be-

---

[*] Ernst Fischer "Global momentum loss in a non expanding universe" *Astrophysics and Space Science*, vol. **190**, No. 1 (April 1992), pp. 149-153; *arxiv*: astro-ph/0805.1638.

cause both Fischer's papers from the early 90s were listed on *arXiv* in 2008. Is this the Age of Aquarius?

Another scientist using laboratory-proven techniques is Jacques Moret-Bailly. He suggests that cosmological redshift may result from the Coherent Raman Effect in Incoherent Light, known to physicists as CREIL. Technically, it describes the transfer of energy between higher and lower frequency light through interaction with the Raman polarisation of the medium. That's quite a mouthful, but it's worth taking a closer look at.

Let's review what we know about coherence in electromagnetic radiation (EMR—a fancy label for light). Without getting clever, we may say that coherent waves have harmonic integrity. A difference in the number of waves per unit of time between emitter and observer represents an increase or decrease in *length*, and is therefore a Doppler shift.

An important consequence of this is that a Doppler-like redshift can occur only in *time-incoherent* light. At the same time, a Doppler-like redshift must be *space-coherent* (that is, have undisturbed wave surfaces) so that the resulting image is not blurred. A further requirement of Moret-Bailly's theory is that the energy drop in the redshifting process should not be quantised, so that the spectra are not blurred.

With these principles on the table, Moret-Bailly set forth. Remember, the exciting, or incident, beams must have the same wave surfaces as the exiting scattered beams. In CREIL experiments, laboratory observations showed that the effect results from a mixture of Rayleigh and Raman scattering, with the exciting and scattered beams having phase resonance.

There was therefore a redshift in the light without blurring of the image. Jacques Moret-Bailly has described an effect that is empirically supported, and not only provides a lucid explanation for the Hubble redshift without invoking an expanding Universe, but is also extremely useful in a number of other areas of astrophysics, including the frequency shifts on the Sun measured by Marmet and Reber.[*]

---

[*] Rayleigh and Raman scattering represent two different types of interactions in electromagnetic radiation. The former is an elastic collision where photons bounce off atoms without losing energy, while the latter is an example of inelastic collisions where energy is lost by the light and as a result it changes frequency (redshifts).

There are literally dozens of carefully constructed, scientifically sound alternatives to the "redshift-means-expansion" hypothesis. Of, course, every single one of them is questionable, and none is capable of direct verification at cosmological scales because we sit remotely at one end of the process.

However—and this is extremely important—the standard explanation itself is not exempt from the same level of sceptical scrutiny. In fact, the conventional model is arguably the least convincing of all, given that it relies on the opportunistic invocation of untested theory. Any of the competing examples mentioned is markedly sounder in the strict scientific sense.

All well and good, but when it comes to physics, I'm hard to please. Some of the theories put forward show enormous promise, but they all have loose ends that leave me unsatisfied. Many are various complexions of pure hocus-pocus. What I'm looking for is a more complete causal description of these redshifts, something that wraps it up more neatly, and covers all the bases. I like to reduce things to first principles and build upon that. All of these mechanisms seem to agree on one fundamental tenet: We are not dealing with a vacuum. The Maxwell-Einstein-de Sitter paradigm seeks to explain the passage of electromagnetic radiation *in a vacuum*, and I suggest that it is precisely there that it goes wrong.

There is no evidence that I'm aware of, that any vacuum exists in nature. The space surrounding us, where we do our measurements, has everywhere some energy density. Interstellar and intergalactic space is nowhere empty; it teems with atoms, ions, molecules, particles, neutrinos, light, all carrying energy, and most presenting mass or mass-energy interactions.

I'm emphasising this point because Maxwell's EMR equations, Special Relativity, and de Sitter space all invoke an abstract ideal not found in the real world. Given that space contains a baryonic medium (irrespective of density) it is impossible for light to pass through and remain pristine. Non-baryonic media are, conveniently for the theory, completely transparent and translucent, so I shall studiously avoid including them here.

I will now list more than 30 examples of redshift mechanisms. I am indebted to my friend and colleague John B Eichler for permission to use the following table from his soon-to-be-published book, An Infinite Universe. It is a summary of a comprehensive review being undertaken by Canadian physicist Louis

Marmet. I have modified Eichler's table to suit the layout of this work.

To review—a suitable redshift mechanism should use known physics, be linear with distance or time, and allow the clear images that we acquire every night. Of course, it should respect the conservation laws, fit with thermodynamics, and studiously avoid Dark Matter, Dark Energy, and other forms of black magic. As a bonus, if it plays a part in presenting the microwave and X-ray backgrounds, so much the better. It's a tall order, I know.

Various proposed redshift mechanisms

Mechanism Description	Name Reference(s)
1. A changing metric of space & time	
Doppler Effect	Hubble (later recanted)
Expanding space, creation of spacetime in voids	Gamow
Lambda-CDM Cosmology inflation	Guth, Wright, Perlmutter
Lobachevsky Space	Von Brzeski
Spatial change of Time Flow	Poliakov
Large Cosmological Constant	Mannheim
Scale Expanding Cosmos	Masreliez
Machian General Relativity	Booth
Quantum Celestial Mechanics Gravitational Potential	Preston, Potter
2. A changing property of matter	
Intrinsic	Narlikar, Hoyle, Arp, Vishwakarma
Velocity Dependent Inertial Induction	Ghosh
3. A changing property of light or an interaction of light with itself	
Finlay-Freundlich Hypothesis	Finlay-Freundlich
Photon Decay	Lewis
Effect of Half-Life	Stolmar
Heisenberg Effect	Caswell
Gravity Nullification Model	Singh
Scalar Potential	Roscoe
4. An interaction between light and matter	
Thomson/Compton Scattering	(known physics effects)
Rayleigh Scattering	(known physics effect)
Gravitational Drag	Zwicky
Atomic Secondary Emission	Paul Marmet
Dispersive Extinction	Wang
Plasma Redshift	Brynjolfsson
Redshift Theorem	Rizzi, Laio
Coherent Raman Effect on Incoherent Light	Moret-Bailly
Electronic Secondary Emission	Ashmore
Wolf Effect	Wolf, James, Roy
Spectral Transfer Redshift	Louis Marmet
Extinction Compton Scattering by Relativistic Electrons	Vaughan
Thermalisation	Gallo
Gravitational Interaction	Mayer
Eternal Contracting Universe	Wilson

Well, to the best of my judgement, we *do* have such an explanation, and it comes courtesy of a stringbean Icelandic particle
physicist named Ari Brynjolfsson. You may not have heard of
him before, but his pedigree is 24 carat. Just look at his citation in
Who's Who in America and you'll see what I mean. Please permit
me to introduce you to a remarkable scientist. Ari Brynjolfsson
left Iceland as a young man to pursue his passion for physics at
the Niels Bohr Institute at the University of Copenhagen. There
he ultimately earned two doctorates—Mag. Sci. (PhD) and Dr.
Phil. (DSc). After some post-graduate work at the University of
Iceland, Brynjolfsson took up a position as Alexander Humboldt
Fellow at the University of Göttingen. At this point he was hired
by none other than Niels Bohr himself to build a radiation centre
in Denmark, the largest such facility in Europe. Ari told me pri-
vately how he arrived at his groundbreaking theory: Around
Christmas 1978, his young children questioned him about Big
Bang Theory, and in contemplating his answer to them arrived at
what he calls Plasma Redshift Cosmology. Much as I prefer to
avoid sweeping, all-embracing models that always seem to answer more than is reasonably asked, most of this idea works for
me, at least, those parts of it that I can understand. It is not always a done deal, extrapolating laboratory evidence to the cosmos, but it's a fine place to start.

To the astrophysicist, plasma is a completely ionised gas-like
fluid of free protons (hydrogen nuclei) and electrons. It is quite
different from anything else we come across in the world, so
much so that it is referred to as the fourth phase of matter, after
solid, liquid, and gas. More importantly, it is the most widely
seen substance in the Universe. Not only is it the part of all incandescent stars that shines, but it is also infused copiously in interstellar and intergalactic space. It would be ludicrous to ignore
the unavoidable interaction between cosmological photons and
intervening plasma.

Ari Brynjolfsson finds no need for bespoke physical theory.
As he puts it,

> The plasma redshift cross-section is deduced from conven-
> tional axioms of physics without any new assumptions. It
> has been overlooked because it is insignificant in ordinary
> laboratory plasmas; but it is important in sparse hot plas-

mas, such as those in the corona of the Sun, stars, quasars, galaxies, and intergalactic space.[*]

The principle is really quite straightforward. Intergalactic light must navigate through vast swathes of hot plasma, and it was discovered fairly recently (and subsequently verified in the lab) that this leads to energy loss, and that, as we now well know, means redshift. The energy lost serves to heat the plasma, thereby addressing some thorny problems in space science, like, why is the Sun's corona hotter than the surface below it? Admittedly, the technical underpinnings do require Quantum Mechanics, but that is standard when dealing with quanta like photons. In the broad view, there is nothing non-standard about Plasma-Redshift Cosmology, and it is really a very clever interpolation of orthodox physical principles. I think Brynjolfsson has a winner here.

Let me conclude with a specific example of assumptions leading to distance anomalies that does not involve quasars. What I would like to do instead is bring to your attention a technique that deals a crippling blow to the idea that redshift indicates distance. The Tully-Fisher Relationship is one of the most solid theories in use on the distance ladder, because it is based upon good physics with an experimentally verified foundation. Compared with the fragile reliability of mooted Standard Candles like Cepheids and Type 1A supernovae, here at last is a method that inspires some confidence. Yet it gives astonishingly anomalous results, when compared with the expectations of expansion theory!

David G. Russell is a New York high school science teacher engaged in an ongoing, novel study of spiral galaxies in the Virgo Cluster, and the Tully-Fisher Relationship is his weapon of choice. TFR, effective for as far as we can measure rotational redshift, describes the empirical relationship between the rotational speed of some classes of spiral galaxies and their luminosity.

It was found by Tully and Fisher in 1977 that certain classes of galaxies displayed signature relationships between rotation and brightness. It's quite logical—the rotational speed of a galaxy depends upon its mass (by Newton's law), and mass in a galaxy

---

[*] Ari Brynjolfsson, "Plasma-Redshift Cosmology: A Review", presented at the 2nd Crisis in Cosmology Conference (CCC2), Port Angeles, WA, 2008. To appear in the proceedings. Please see also Ari Brynjolfsson, "Redshift of photons penetrating a hot plasma" *arXiv*: astro-ph/0401420.

is proportional to the number of stars it contains. More stars means brighter light, so we would expect to find a direct correlation between the speed at which the galaxy rotates and its true luminosity. Once again, with the inverse-square law for the dimming of light, we can use the difference between true brightness and perceived brightness to calculate distance.

His purpose was not to challenge redshift distances, but rather to use them as part of a method to define the physical boundaries of the cluster. However, in a surprising turn of events, Russell discovered that there was a wild spread of redshifts for spirals in the cluster, and that just didn't stack up against the Hubble law.

It was quite astonishing: Once he had identified and confirmed which galaxies were physically bound to the cluster, within its gravitational sphere, he set out a table containing other properties of the cluster galaxies, including redshift. He discovered, quite incredibly, that the swing in redshifts was remarkable, and even more worrisome for expansion theory, that redshift values were bracketed by galaxy type![*] Expressed as recessional velocity, that would mean that (in Russell's words)

> ...giant Sb galaxies are approaching the Milky Way with a mean velocity of −898 km s[-1] while the giant ScI galaxies are receding from the Milky Way with a mean velocity of +824 km s[-1].[†]

I can imagine a humorous scenario: A graduate student who was not completely comatose sat up, rubbed her eyes, and said, *"Hey, there's something horribly wrong here!"* Dave Russell had systematically (and independently of Halton Arp) discovered that celestial objects confirmed to be in the same geographical neighbourhood could have significantly different redshifts.

He didn't leave it there. In 2009 he published The Ks-band Tully–Fisher Relation – A Determination of the Hubble Parameter from 218 ScI Galaxies and 16 Galaxy Clusters in the Journal of Astrophysics and Astronomy, and the orthodox view of a distance modulus based on cosmological redshift received yet another body blow. The measurements Russell obtained from these 200-

---

[*] Redshift by galaxy type lends weight to the idea that redshift is at least partially a function of internal energy.
[†] David G. Russell, "Intrinsic Redshifts and the Tully-Fisher Distance Scale" *arxiv*: astro-ph/0503440.

odd objects returned a Hubble expansion rate H_0 of 84 km per second. This is a definite problem for the L-CDM model. He broke the news to me in an email:

> "In this article I find $H_0 = 84$ which creates either an age crisis or a conflict with the matter density of the universe. It is very difficult to reconcile $H_0 = 84$ with lambda-CDM cosmology. I also discuss potential flaws with the Hubble Key Project final value of $H_0 = 72$.[*]

Dave Russell has provided a new angle on the empirical confirmation of intrinsic redshift. No doubt he will attract the wrath of the party-line establishment for his troubles, and like Chip Arp forty years before him, go from hero to zero amongst those blessed with the job of allocating observatory time and pages in mainstream journals. But the empirically solid results he leaves behind are a monument to the power of observation to falsify popular theory. It is an ominous crack in the prevailing paradigm's wall.

The idea of distance-dependent redshift has become an extremely important quantity in astrophysics, and supports a large body of theory. In cosmology, it gives us radial calibration along line-of-sight that determines almost exclusively the depth in 3-D representation of structure, and what's more, almost single-handedly defines the phenomenon of expansion. By rights, though, it should be no such thing. As a premise, it is utterly false and substantiated by nothing at all. As we have seen, it is a pretender to the throne, lacking any real pedigree. I don't think it's anywhere near as important as it makes out, for the simple reason that it doesn't present itself in observation, and in any case, even if we were to see it somewhere, we wouldn't know what exactly to make of it.

Isn't it strange how energetically devoted theoreticians tackle the problem of taming anomalies? With whips, trowels, and dollops of fudge, it seems they can get any wayward, prodigal observations back into the fold, and without raising a sweat, realign dissident interpretations with their preferred model. In their hands, square pegs and round holes are raised to an art form. Despite the clear warning of Hubble himself, astronomers suc-

---

[*] David G Russell, "The Ks-band Tully–Fisher Relation – A Determination of the Hubble Parameter from 218 ScI Galaxies and 16 Galaxy Clusters" *J. Astrophys. Astr.* **30** (2009), 93–118; *arxiv*: astro-ph/0812.1288).

cumbed to the urgent need for a way to establish remoteness in space of celestial objects, and the even sexier imperative to drive the exciting new expanding-universe model forward. Consequently, the redshift-gives-distance idea was carved into the wall of astrophysical law.

From an independent, objective point of view, the accepted explanation doesn't even get to first base. Let me say it again — the only reason, and I mean the *only* reason that such a blatantly improbable theory has seen the light of day is because the Standard Model requires it. Redshift, like luminosity, has no fixed relationship with velocity, and we should not forget these things in our enthusiasm to explain everything in sight. Whatever we decide to do, it is clear at the very least, that redshift is not demonstrably proportional to distance — or recessional velocity — in all cases, and indeed, has been reported to be correlated with luminosity for galaxies in local, non-expanding space.

The Hubble Law and attendant redshift-based expansion are a myth. We might even dare, in this age of Al Gore's pseudoscience, to call it a Convenient Untruth.

> *Discussion: Cosmological redshift is a cocktail, not a single ingredient. It is currently quite impossible to determine the original degree of redshift in cosmological objects, and we therefore cannot claim to know or infer the degree of change occurring during astrophysical processes.*

Chapter 5

Quasars

Traffic lights or milestones?

They are assumed to be the brightest objects known, but observations show they might well be exceptions that demolish the redshift rule. "There is always something rather absurd about the past."

(Sir Max Beerbohm).

"I don't see the logic of rejecting data just because they seem incredible."

(Sir Fred Hoyle).

♦ ♦ ♦ ♦ ♦ ♦ ❖ ♦ ♦ ♦ ♦ ♦ ♦

The fly in the ointment of modern redshift astrophysics is undoubtedly the inconsistent behaviour of Quasi Stellar Radio Objects. If seen without theoretical prejudice, quasars present an alarming challenge to orthodox cosmology, and were it not for the relentless courage of a few dedicated astronomers, we might never have known about it.

In 1963, Alan Sandage and Thomas Matthews combined optical and radio astronomy to identify quasars for the first time. They initially confused them with massive stars, hence the intriguing name "quasi stellar object." Right off the bat, these creatures of the cosmic night were noticeably peculiar, and moreover, were actually defined by their strangeness. Most importantly, they displayed redshifts significantly higher than other objects seen on the sky, and that led to some really far-fetched conclusions. Quasars created difficulties for physical theory because at their redshift-implied remoteness, they would by known physics be impossibly bright. Quasars are thought to be very compact objects, typically only about one light year across. If they really were that far away, they would be so energetic that their luminosity enters the realm of metaphysics.

Everything those pioneers *thought* they knew about quasars depended crucially upon distance. If they were highly redshifted (and that they were, no doubt) then they were spectacularly far away. There was nothing more remote ever seen on the whole wide sky, and that, in turn, meant they were astrophysical superpowers. Their extreme distance implied impossibly high levels of intrinsic brightness—by a factor of 10,000, no less! Astrophysicist Howard Yee computed in 1988 that the newly-discovered Einstein Cross would require for its projected redshift-indicated luminosity a staggering 100 *billion* solar masses within the small space occupied by the quasars. That's 1,000 times more massive than the Milky Way's nucleus. I'm not kidding! Even more onerous was the precision measurement of radial expansion rate by very long baseline radio interferometry. Quasars appeared to be expanding at up to ten times the speed of light, with obviously serious implications for underlying theory and Einsteinian physics. All these fantastic properties were a direct consequence of their being so incredibly far away, and that, of course was given indisputably by their redshift.

We can't argue with one thing: Quasars *are* peculiar objects, whichever way we look at it. They have fearsome reputations (mostly undeserved), and behave strangely. So strangely, in fact, that they suggest by their very dissidence that they may hold secrets that could unravel some incredibly puzzling problems relating to where we all came from. They certainly deserve to be studied by every astronomer at some stage of his career.

Cosmology in the modern era is never slow to grasp a transient opportunity, even if it means adjusting the facts a wee bit. Quasars were quickly utilised to shore up the faltering theory, and rather ironically, soon became the regimental colours of the Big Bang brigade. Cambridge astrophysicist and Astronomer-Royal-to-be Martin Rees wove them into an intriguing theory about deep space, and the applause was resonant. Meanwhile, Halton Arp pointed the gargantuan 200-inch Palomar telescope at

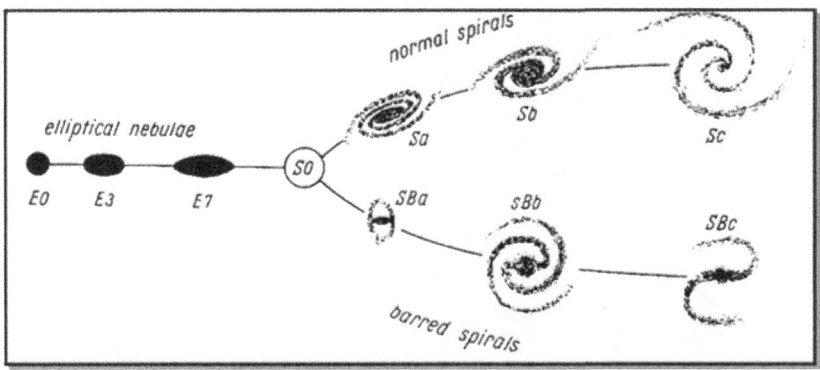

Figure 14: Hubble's "Tuning Fork" galaxy shape classification, as displayed in The Realm of the Nebulae.

the night sky and took photographs. That, my friends, was when this whole sorry tale turned hostile.

Halton "Chip" Arp is a gentleman and a scholar; an honourable man with clear ethics and firm principles. He is also one of the finest observational astronomers of our era. I hold him in the highest esteem. Surely, you will understand that I am not drowning in admiration for those pompous idiots who shut the door in his face, and attempted a heavy handed suppression of the photographs he had taken of celestial objects. There hangs a tale of intrigue and deception here, and I am going to tell it to anyone who cares to listen.

It would be fair to say that the controversy surrounding quasars and the implied phenomenon of intrinsic redshift may be attributed mainly to the early observational work of Halton Arp. His interest in the astronomical distance ladder, stemming from his doctoral work with Edwin Hubble and subsequent 2-year stint observing Cepheids in South Africa, brought redshift into focus.

In 1965, two oddities caught his interest: Galaxies appeared to be in turmoil, showing signs of great internal stress and presenting themselves in ways that could not neatly be accommodated on Hubble's "Tuning Fork" galaxy scheme; and an unusual prevalence of quasars, in pairs or more, aligned closely across active (Seyfert-like) galaxies. Sandage had collaborated with Gerard de Vaucouleurs in 1958 to try (unsuccessfully) to systematically accommodate the wildly varying structural types of galaxies, and

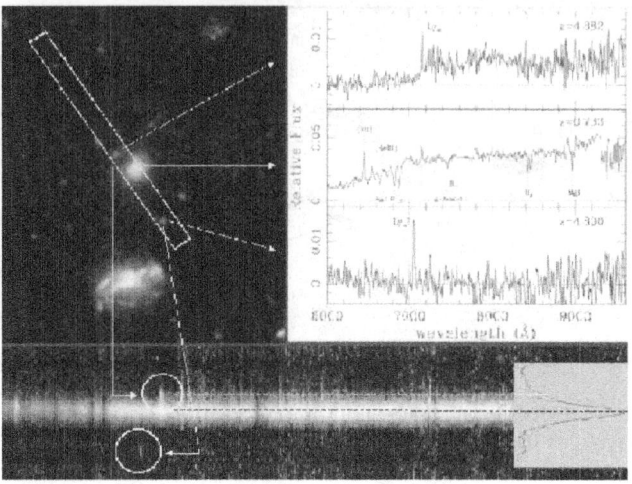

Figure 15: Quasars of z = 4.882 and z = 4.800 are aligned across an emission line galaxy of z = 0.733. Lyman alpha emission can be seen intruding into the spectrum of the lower redshift, central galaxy. (Vanzella et al. astro-ph/0406591) (Image courtesy of Halton Arp).

in 1966, Arp published a collection of these images in his classic Atlas of Peculiar Galaxies.

The furore that followed split the astrophysical community, with most astronomers declaring that close alignment of quasars with AGN was just chance, line-of-sight coincidence with no statistical or physical significance. A small minority took an alternative view, however, amongst them (besides Arp) Margaret and Geoff Burbidge, Fred Hoyle, Jayant Narlikar, and Jack Sulentic. Arp found some rather odd angular associations on the sky between Seyfert (that is, *active*) galaxies and pairs of quasars aligned with their polar axes. He took photographs of these systems, collated them in a paper, and submitted it for publication. The editor of the Astrophysical Journal at that time was the eminent and often opaque theorist Subramanyan Chandrasekhar, and he reacted with contempt.

In a brutal, scandalous crackdown that has tainted and besmirched the gentle art of astronomy to this day, the powers-that-be closed ranks. How *dare* an astronomer attempt to publish photographs of unorthodox objects? They forbade Arp access to the major West Coast observatories, and put what they obviously hoped would be an unbreakable choke-hold on his career. Fortunately, not everyone in the world of astronomy was as myopic as those unfortunate individuals who then controlled telescope time and publication in the United States of America.

After his banning in the early 1980s, fate intervened. Arp took up employment at the *Max Planck Institut für Extraterrestrische Physik* (MPE) in Munich, where he was able to continue acquiring images in X-ray of objects he had previously observed optically. Ironically, the enforced migration from optical to X-ray dealt Arp an unexpected trump—previously unseen linking structures were thereby revealed, and the great value of composite images in various wavebands was obvious. The MPE's cutting-edge X-ray telescope, says Arp,

> ...picked out the most energetic objects with ease, and the telescope was still small enough so that it had sufficiently large field to include the crucial objects which were related to the central progenitor galaxies.[*]

Those seeking to suppress his research had shot themselves squarely in the foot.

The official response to Arp's burgeoning observational evidence was that it was just rare coincidence, so the first question that needs to be answered in a review of anomalous quasar data is, "What is the statistical significance of the samples being cited?" Put another way; are anomalous quasar associations not in fact just extremely uncommon events that can be written off to chance alignments and optical illusion?

Arp was the first astronomer to compile a catalogue of strange galaxies, and it became a classic. His first volume, *The Atlas of Peculiar Galaxies*, originally a supplement to the *Astrophysical Journal (ApJ)*, is currently out of print, so I reference here Kanipe and Webb's version, which contains all the images. It lists 338 disturbed galaxies. They are known as the Arp galaxies, and have Arp numbers from 1 to 338 in the order presented in the atlas.[†] Up to then, the samples available to him had been limited in scope, but contemporary large-scale cosmic surveys, prominently the *Sloan Digital Sky Survey* (SDSS), immediately introduced mil-

---

[*] Halton Arp, Seeing Red: Redshifts, Cosmology, and Academic Science (Montreal: Apeiron, 1998).

[†] Jeff Kanipe and Dennis Webb, The Arp Atlas of Peculiar Galaxies – A Chronicle and Observer's Guide (Richmond, VA: Willmann-Bell, 2006).

lions of objects to the field of study. Amongst them were more than 40 000 positively identified quasars.

The two deep field surveys are also invaluable sources of redshift data. The 2dF Galaxy Redshift Survey (2dFGRS) lists around 250,000 galaxies, and the 2dF Quasar Redshift Survey (2QZ) examines more than 25 000 quasars. There is now no shortage of material for us to analyse, and the answers coming back from these surveys do Big Bang Theory no favours at all. Arp's subsequent publications continued to display observational evidence of these associations, now improved by advanced instrumentation to include more detail than just tight angular spread, and led ultimately to his Catalogue of Discordant Redshift Associations, published in 2003.[*]

It takes only one verified discordant measurement to falsify a theory. That's all. Throughout this book you will find references to Mike Disney of Cardiff University. All show that I am probably more in agreement with Disney than with anyone else in the field of astrophysics. There is just one discord, however. At the first Crisis in Cosmology Conference (CCC1) in Portugal in 2005, he made rather wry comment in response to reference to the quasar observations of Halton Arp. "Arp!" he said grumpily, "He sees a one-legged man and concludes that all men are one-legged!"

I'm afraid you are wrong on two counts, Dr Disney. Firstly, Arp's examples of one-legged men (AGN-linked quasars) run into tens of thousands. Secondly, it takes just *one* confirmed observation of a one-legged man to refute the contention that *all* men are two-legged. The observations are solid.

In fact, quasars are *so* one-legged it's quite amazing that anyone would even attempt to accommodate them in the Standard Model, yet that is exactly what some scientists are trying to do. It's a hopeless cause. Perhaps it would be useful to refer to a 2005 presentation by Michael Strauss of the Space Telescope Science Institute (STScI) and Princeton University, entitled "Active Galaxies at Low and High Redshift: Type II Quasars, Reionization, and Other Insights from the Sloan Digital Sky Survey." In this presentation, Strauss makes the following points in reference to quasars with $z \cong 6.5$:

- There was insufficient time for so much mass to accrete;

---

[*] Halton Arp, *Catalogue of Discordant Redshift Associations* (Montreal: Apeiron, 2003).

- There was insufficient time for quasar metallicity to reach the observed level;
- SDSS finds no evolution in the metallicities of quasars with redshift; and
- Higher-redshift quasars would stand a much greater chance of being lensed, yet none of the $z = 5.7 - 6.5$ quasars in the SDSS survey is lensed.

The physical association of high and low redshift objects is ubiquitous, and the discovery of binary stars with different individual average redshifts sets a startling precedent. Binary stars interact visibly with each other at very close range. There can be no doubt that they form a single system and at the scale of stellar distances, are both equally far from us. As usual, this evidence is ignored, probably in the hope that it will be slowly buried under the sands of time.

In the documentary programme Universe—the Cosmology Quest, Geoffrey Burbidge is succinct:

> If you see two objects close together with very different redshifts, you only have one of two explanations. One is that a large part of the redshift has nothing to do with distance. The other is that it's an accident. *So the real issue...is how frequently do you expect to see accidents?*[*] (Emphasis is mine.)

Morley Bell, of the Herzberg Institute of Astrophysics in Canada, sums it up thus:

> Because the belief that the redshift of quasars is cosmological has become so entrenched, and the consequences now of it being wrong are so enormous, astronomers are very reluctant to consider other possibilities. However, there is increasing evidence that some galaxies may form around compact, seed objects ejected with a large intrinsic redshift component from the nuclei of mature active galaxies.[†]

---

[*] Randall Meyers (Producer), *Universe – the Cosmology Quest* (Floating World Films, 2000, www.universefilm.com) is a science motion picture documentary on DVD, featuring interviews with Fred Hoyle, Chip Arp, Margaret and Geoff Burbidge, Jayant Narlikar, Eric Lerner, and Jack Sulentic, amongst other eminent researchers in the field.

[†] M. B. Bell, "Further Evidence that the Redshifts of AGN Galaxies May Contain Intrinsic Components" *arxiv*: astro-ph/0603169.

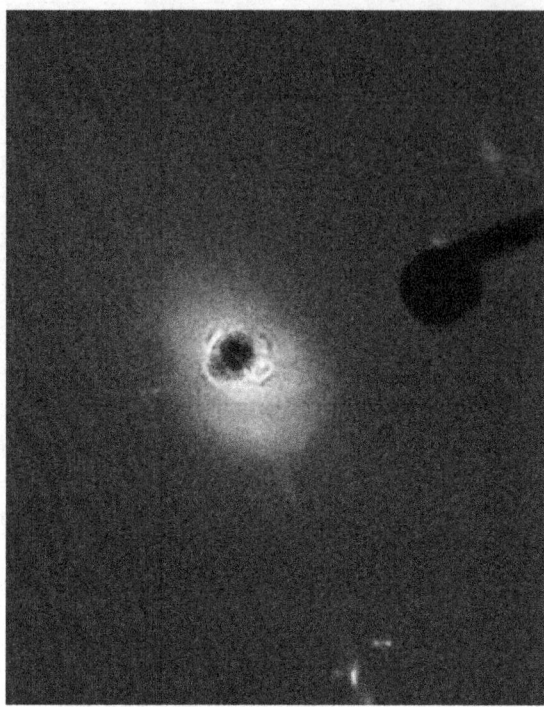

Figure 16: Quasar 3C 273. How far? How big? How bright is it really? Is it a Black Hole? (Image courtesy of NASA, ESA, and STScI).

For quasars to be so easily visible over such immense distances, they must of course be exceptionally bright—not trivially so, but by a factor of 10,000 brighter than any other measured luminescence! Nothing known to man had such intense energy; redshift indicated that these mysterious objects lay at the very boundary of the known universe, yet they were as clearly visible here on Earth as galaxies that lay orders of magnitude closer.

Insisting that they are extremely far away, and therefore exceptionally luminous, can bring some pretty stiff problems to the table. Quasars have been uncovered recently with phenomenally high redshifts, and we would expect them intrinsically to be proportionately brighter on a par, say, with the quasars in the Sloan Digital Sky Survey (SDSS). Unfortunately, they are not. They are miles adrift in luminosity. Chris Willott of the renowned Herzberg Institute of Astrophysics in Canada teamed up with several other astronomers to analyse data from collaboration between Canada and France. It is a definitive study, published under the title "Six More Quasars at Redshift 6 Discovered by the Canada-

France High-z Quasar Survey."[*] A redshift z ~6 is exceptionally high, and puts the quasars in question right at the limits of the observed universe, if redshifts are to be believed.

However, it just doesn't add up. Willott explains: "The new quasars have luminosities 10 to 75 times lower than the most luminous SDSS quasars at this redshift." The quasars are so dull, in fact, that they can't be detected in X-ray, which they are alleged to give out copiously. Oh yes. Wait for it. This is where they throw in Black Holes.

Scientists attempting to justify the remoteness of quasars utilise the hypothesis of "Black Hole Mass." It is Black Holes, they tell us, which provide energy levels quite impossible in our understanding of conventional physics. Without apology, I reject this explanation out of hand as paranormal metaphysics. They might just as well say that the QSO luminosity is caused by green elephants.

The standard version of Black Hole theory is probably best understood (if so generous a verb is appropriate here) in the definitive 1998 essay "Black Holes: A General Introduction" by Jean-Pierre Luminet of the *Observatoire de Paris-Meudon*.[†] It formed part of the reference work "Black Holes: Theory and Observation", and I should suppose we are expected to take it seriously. So that we could easily do so, the author eschews green elephants altogether, and gives us a quaint but rather disturbing story about butterflies.

"Once upon a time," Luminet tells us, "...the butterflies organized a summer school devoted to the great mystery of the flame. Many discussed about models but nobody could convincingly explain the puzzle. Then a bold butterfly enlisted as a volunteer to get a real experience with the flame."

Two butterflies in turn failed to discover anything useful about a nearby candle flame, but a third intrepid fellow did, by Jove! Instead of merely observing the flame as his two more timid brethren did, this chap flew into the flame and burnt himself to a frazzle. The wise chairman butterfly, "...who had observed the

---

[*] Chris J Willott *et al.*, "Six More Quasars at Redshift 6 Discovered by the Canada-France High-z Quasar Survey" *arxiv*: astro-ph/0901.0565.

[†] Jean-Pierre Luminet "Black Holes: An Introduction" *ariv*: astro-ph/9801252 in: F. Hehl, C. Kiefer, R. Metzler, Eds., *Black Holes : Theory and Observation*, (Berlin: Springer Verlag, 1998).

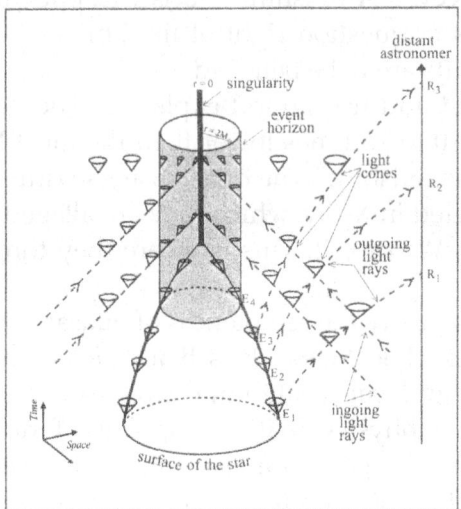

Figure 17: What, no but-
terflies? Fig. four from Luminet's paper "Black
Holes: An Introduction."

action, said to the others: 'Well, our friend has learned everything
about the flame. But only him (*sic*) can know, and that's all.'"

Thank you, Dr Luminet. Now we get it: Better dead and wise
than still breathing while we reject the reality of Black Holes.

The Standard Model tells us that quasars are really Black
Holes located at the centres of galaxies. Black Holes are thought
to surround themselves with an intensely bright halo by vora-
ciously attracting and devouring neighbouring stars. At the National Symposium of the Astronomical Society of Southern Africa
in 2008, I presented a paper that summarised the many anomalous redshift results obtained from direct observation. In it, I dealt
with the question of almost impossible levels of energy needed to
fire quasars up to the suggested levels of intrinsic brightness.

In retrospect, perhaps I shouldn't have used the term "metaphysical." It deeply offended a certain well known astronomer in
the audience, but that was definitely not my intention. The out-
raged professor took issue with me afterwards, stating that he
had been at a recent conference where it was shown (to the good
professor's instant satisfaction, apparently) that the calculated
masses of quasars were easily adequate to drive the mooted luminosity. Well, I don't agree, and upon sober reflection, feel my
use of the word metaphysical was apt. You see, the way that
masses are calculated in these special cases is done by invoking
Black Holes. Normal, common-or-garden physics just cannot do

$$T = \frac{hc^3}{16\pi^2 GMk}$$

Figure 18: The Hawking Black Hole formula. *T* is temperature and M in the divisor is mass. That means that temperature of a Black Hole is inversely proportional to mass—as it radiates away into space (in the Big Bang, for example) it gets rapidly hotter! This is quite the opposite of BBT, which sees the Universe cooling with expansion.

it, so, in order to retain the model, dark and mysterious creatures are brought into play.

The fact is, we do not need them, or indeed even Big Bang Theory, to explain anything we see. The only reason we entertain such outlandish metaphysical concepts is to prop up a cosmological model that really has no support from observation. We do not *see* an expanding universe, we do not *see* curved space-time, we do not *see* systematic, open-ended evolution of structure with time, we do not *see* Dark Matter and Dark Energy said to comprise over 90% of the universe, and we certainly do not *see* Black Holes in any form, shape, or size.

These things were conjured up in mathematical daydreams. The physics of Black Holes is so insubstantial that practically any observed energetic effect can be attributed to them, including concentrations of hyper-gravity, gamma- and X-ray bursts and the spurious super-intense luminosity of quasars.

Stephen Hawking has had a field day with theoretical excess concerning Black Holes. To his credit, he always confesses when he is shown by his own further meditations to be wrong, but then he goes on to make even more absurd conjectural predictions. It's a mind-game, mildly amusing in idle moments, but it really shouldn't be taken seriously as an explanation pertaining to the real world.

It is important to point out that Black Holes have never been observed, not even by eclipsing. No jet black disk has ever been spotted, nor stark silhouette resolved. The radiation signatures attributed to these incredible ogres are at once ambiguous and by definition not from the Black Holes themselves. All the physical "evidence" can be more properly explained by substituting high-density, natural compact objects. Furthermore, the concept itself is untenable in physics; they simply cannot exist, in theory or practice, in reality as we know it. Stephen Crothers and Angelo Loinger have produced compelling evidence in rigorous analysis of the theoretical base in General Relativity that despite claims to the contrary, Albert Einstein was right after all—Black Holes can neither emerge theoretically from Relativity nor can they by any known physical process be brought into existence. Black Holes belong in the mythology of mathematical wanderlust, not in the guileless science of astronomy.[*]

Halton Arp and his colleagues found observationally that three aspects of quasar distribution were anomalous: Their occurrence amongst other objects showed an inordinate prevalence of quasars paired in close (angular) proximity across Active Galactic Nuclei; objects apparently physically associated in space had significantly varying redshifts; and the asymmetrical concentrations of isophotes[†] on AGN/quasar maps indicated that the quasars were moving away from the AGN, confirming Arp's suggestion of ejection.

Arp has to date published four volumes besides his many papers and articles, three in book form. All are in effect catalogues of his observations, and they contain hundreds of examples. It is interesting to note Arp's use of the collective noun "family" in his recent work; it emphasises the important increase in power and resolution of modern surveys. From the first tentative observed alignments of pairs of quasars in the 1960s, we are now introduced to groups of ten or more closely gathered around active galaxies.

The observational evidence supporting Halton Arp's quasar associations is immense, far too voluminous to include here. Please refer to my published summary "A Review of Anomalous

---

[*] Please read these two papers for rigorous falsification of Black Hole theory and practice: Angelo Loinger, "The black holes are fictive objects" *arxiv*: astro-ph/9810167, and Stephen Crothers, "A Brief History of Black Holes" *Progress in Physics*, April 2006.

[†] Isophotes in astrophysics are analogous with isobars in meteorology, or contour lines on relief maps. They connect points of equal magnitude (brightness) on maps of astrophysical systems, and are invaluable in tracing physical connections between objects.

Figure 19: The originally published photograph of active
galaxy NGC 4319 ejecting
quasar Markarian 205, which
shows no bridge to the parent galaxy. (Image courtesy
of NASA, ESA, and STScl).

Redshift Data" for greater detail and more than 50 references.
Pardon me for including the somewhat technical illustration
(above) and quote (below, from Halton Arp's website). There are
many, many such examples now formally declared, and all reach
a clear, common, and surely irrefutable conclusion: Quasars are
physically associated with active parent galaxies at significantly
lower spectral redshifts than the quasars themselves, and are
strongly indicated by the data to have been ejected by them.

> In the Great Observatories Origins Deep Survey, 243 red-
> shifts of objects fainter than 25.5 mag. were observed. Re-
> markably, two of them turned out to be very high redshift
> at $z = 4.800$ and $z = 4.882$. Even more remarkably these two
> fell only 3 and 1.5 arcsecs on either side of an emission line
> galaxy of $z = 0.733$... The picture shown below is probably
> sufficient to convince most people that this is another pair
> of ejected, intrinsic redshift quasars.

Let's take one of the most notorious examples. Of particular
interest to astronomers was the apparent existence of links, like
tails or matter bridges, joining objects anomalously. The classic

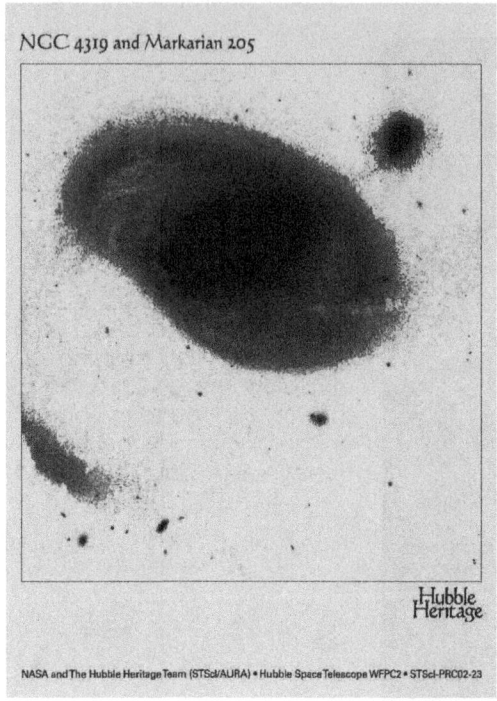

NGC 4319 and Markarian 205

Figure 20: This is a negative image of figure 19, created by Jack Sulentic. The material bridge linking the two is clear.

Hubble
Heritage

NASA and The Hubble Heritage Team (STScI/AURA) • Hubble Space Telescope WFPC2 • STScI-PRC02-23

case, featured on the covers of all Arp's books, is the famous "invisible" bridge linking NGC 4319 and the quasar Mrk 205. Halton Arp tells the story:

> In 1971 with the 5 metre telescope on Mt. Palomar a luminous bridge was discovered between the low redshift galaxy NGC 4319 and the much higher redshift quasar, Markarian 205. Because this contradicted the assumption that redshift was invariably a measure of velocity and distance, it invalidated the hypothesis of an expanding universe. Conventional astronomers fiercely resisted this evidence but as it accumulated for this and numerous other similar examples the results were increasingly suppressed and ignored.

In the early 1980s, Jack Sulentic soundly debunked two much-cited papers that claimed the observed bridge simply did not exist, and in 2007, he reacted again to similar claims, this time in a press release from Hubble Heritage. The picture released did not show a bridge, whereas in the original image captured by the space telescope, it is clearly there. The picture had been presented in a peculiar light indeed. Wonder why?

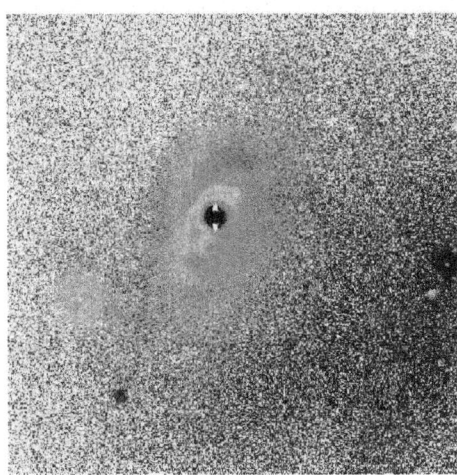

Figure 21: The low-redshift active galaxy NGC 7603, with the clearly visible tail extending to the higher-redshift quasar NGC 7603B to the south-west. (Image by kind permission of Martín López-Corredoira).

In the unadulterated HST image, Sulentic says,

> You can see the narrow core in the connection, which HST is able to detect because of its excellent resolution. It is seen exactly where we found it in the earlier studies...Hubble Space Telescope has in fact, confirmed our earlier work.

In conversation, Margaret Burbidge was puzzled by this approach, often by scientists who seemed in other respects quite reasonable. "I could understand it if they said that the bridge was impossible to explain," she remarked quite downheartedly, "but to simply say that it's not there, to deny its very existence when it is clearly seen in the photograph, is just incomprehensible." Indeed it is, Dr Burbidge.

Then there's the problem of trying to get the evidence published. Believe me, this is not an imagined problem; it's very real. The principle that science progresses by discovery of the unexpected and unexplained seems lost. At a meeting of the American Astronomical Society held in Texas in 2004, Margaret Burbidge presented a paper that she had co-authored with Arp and several other leading astronomers. It detailed the discovery of a high redshift quasar close to a low redshift galaxy, and it was extremely significant. In fact, it should in the absence of undue official interference have rung the death knell of redshift-based expansion. You see, the high redshift quasar lay *in front of* the galaxy NGC 7319! In it, the authors presented observational evidence that a strong X-ray source with relatively high redshift lay in the *fore-*

ground of NGC 7319, an active galaxy with relatively low red-shift.[*]

Several tests were conducted to determine whether or not it lay in the foreground, for if it were, beyond reasonable doubt, the case would be conclusive. Is the QSO behind the galaxy?

> One obvious question suggests itself, namely: Does the colour of the QSO indicate that it is inordinately reddened and therefore obscured as if it were a background object? We find that it is about 0.1 to 0.2 mag. bluer than average.

All the evidence points to a foreground quasar. The astonished audience, among them the most prominent astronomers and cosmologists of the time, was stunned to silence. "It was incredible," Chip Arp told me afterwards, "in that huge gathering, you could have heard a pin drop."

However, when time came to seek publication of the paper in the Astrophysical Journal, it was made conditional upon the inclusion of a contrived counter-argument that the quasar lay *behind* the active galaxy. The authors had no option but to comply. It was a complete and utter travesty of all we hold dear in the ethics of science.

Nor is it just Halton Arp who finds evidence supporting these associations. Martín López-Corredoira and his colleague Carlos Gutiérrez have tried, with only marginal success, to get telescope time at their observatory for a project that could swing the argument either way, *depending on what is seen*. Despite these restrictions, López-Corredoira persevered in the face of severe institutional resistance, and produced volumes of observational evidence of systems that are, at the very least, peculiar. Some of his images of quasars embedded in the trailing arms of spiral Seyfert galaxies are nearly impossible to deny. That doesn't stop his opponents, when confronted with a visible matter bridge, from simply insisting, "It doesn't exist."

One of the most explicit examples of redshift anomalies in active galaxy-quasar associations is that of the NGC 7603 system studied by López-Corredoira and Gutiérrez. The central, progenitor object is a Seyfert galaxy with a redshift value $z = 0.029$. It is

---

[*] Pasquale Galianni, E. M. Burbidge, H. Arp, V. Junkkarinen, G. Burbidge, Stefano Zibetti, "The Discovery of a High Redshift X-Ray Emitting QSO Very Close to the Nucleus of NGC 7319" *arxiv*: astro-ph/0409215. The amended paper was subsequently published in the Astrophysical Journal.

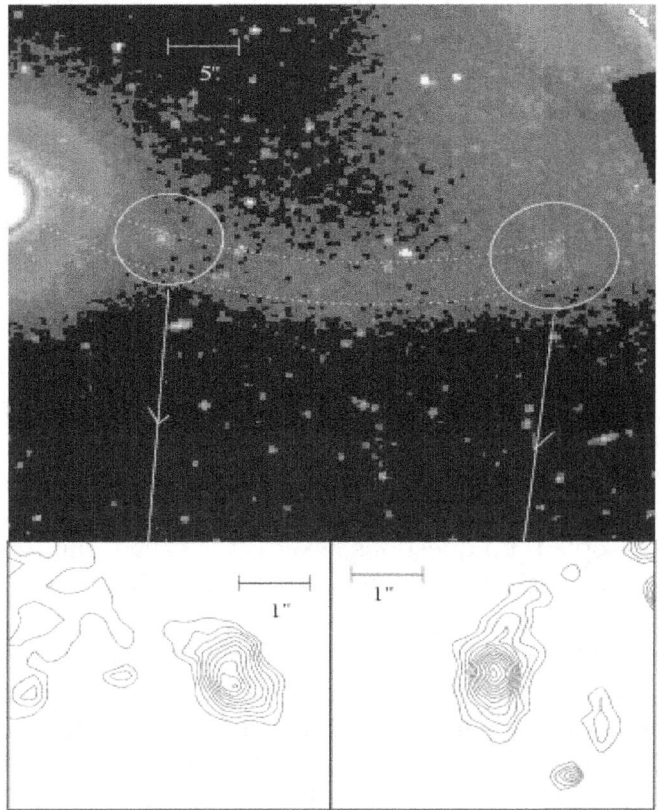

Figure 22: A more detailed image of the tail in the NGC 7603 AGN-QSO system. Two more high-redshift emission galaxies (probably quasars) are clearly embedded in the tail. Significantly, the redshift values in the tail diminish in the direction of NGC 7603B. The contour lines in the two diagrams at the bottom are isophotes, joining points of equal radiant intensity. (Image by kind permission of Martín López-Corredoira).

linked by a tail following a classic Fibonacci curve to the quasar NGC 7603B, with redshift $z = 0.057$. Don't be fooled by the decimals—the redshift z-value of NGC 7603B is twice as high as that of the Seyfert it is physically attached to. That's bad enough, but it gets worse. López-Corredoira and Gutiérrez examined the bridging material closely and found two emission galaxies (probably quasars, but that's not important right now) embedded in it.

I harbour a grave suspicion that the assumption of embedding follows two distinct sets of rules, depending on whether it supports or challenges the Standard Model. If it were not for their redshift values, these two intervening quasars would be taken without question to be embedded objects; in this case, they are spurned as line-of-sight coincidences. This is only because red-shift-distance is unquestioningly assumed.

Careful analysis of Hubble Space telescope data strongly supported the view not only that these two mini-galaxies were physically part of the tail, but that the parent galaxy was distorted and highly active. All the signs pointed to ejection. The glaring problem for standard theory was that these two interlopers were also at much higher redshift than their progenitor galaxy, $z = 0.245$ and $z = 0.394$, respectively. This association is quite impossible, say the scribes of the Standard Model, yet there it is, as clear as daylight.

I would bet a pound to a pinch of salt that if such an association were representative of the Standard Model, it would be instantly accepted, no questions asked. It is called into doubt—or blissfully ignored—simply because it is not what consensus cosmology wants to see, or wants us to see.

Most of the celestial objects we are able to study appear static on the sky, as if they were frozen in time. This is an illusion; the Universe is indeed internally dynamic, with things whizzing around in all directions, towards each other and away. They appear to be standing still because the vast time and distance scales separating us, and them from each other, damp out the visual effect. However, for objects that are relatively nearby, we can detect the movement of one from another, and this is called "proper motion". The greatest example of proper motion is Barnard's Star, which has moved appreciably in relation to its neighbours in the 30-odd years that I have been looking at it. It migrates 10.3 seconds of arc per year, which equates to the angular breadth of the full Moon in less than 200 years.

The point I wish to emphasise is this: We can see and measure the drift of Barnard's star *only* because it is so nearby. In fact, at a trifling 5.97 light years distant, it is second-nearest to our Solar System after the Alpha Centauri triplet. If it were a million light years away, or even a thousand, we would have great difficulty in detecting the proper motion of the star. Whilst it is not quantitatively exact by any means, proper motion is nevertheless a good indicator that the object under review cannot be very far. What I am going to tell you now is hardly ever mentioned in polite society, at least when orthodox cosmologists are present. It is

Luyten	Burbidge	Ap Mag V	Proper motion	RA (2000.0)	DEC	Z
LB 8956	0854+191	17.6	60.8±18	8 57 26.85	18 55 24.4	1.896
TON 202	1425+267	15.68	52.6±16	14 27 35.63	26 32 14.7	0.362
LB 8991	0855+188	17.3	50.5±18	8 58 30.12	18 37 9.0	1.013
PHL 1033	0131+037	18.7	48.8±13	1 33 43.34	3 57 35.2	0.255
LB 9029	0856+189	17.7	36.1±18	8 59 27.44	18 43 49.8	1.286
Br 337	1305+352	17.62	35.0±16	13 7 49.61	35 1 41.4	0.30
PHL 1106	0139+059	18.3	32.2±13	1 41 59.52	6 12 3.6	0.345
LB 9388	0906+167	17.2	30.9±18	9 9 16.10	16 35 23.6	1.07
LB 9013	0856+170	17.4	29.1±18	8 58 52.40	16 51 28.5	1.454
PHL 1072	0135+056	18.3	28.6±13	1 37 48.81	5 55 26.2	0.615
PHL 1070	0134+033	17.6	28.4±13	1 34 48.0	3 21 00	0.079
PHL 1194	0148+090	17.83	26.2±13	1 51 30.88	9 17 25.4	0.299
PHL 3632	0139+061	17.80	25.1±18	1 42 34.65	6 25 39.4	1.479
PHL 8462	0237-233	16.63	23.6±13	2 40 8.11	-23 9 18.0	2.225
PHL 1119	0140+081	17.1	23.3±13	1 42 54.69	8 22 12.2	0.119
LB 9707	1523+214	17.96	22.1±18	15 25 22.33	21 14 6.7	1.924
TON 616	1223+252	16	21.9±10	12 25 39.55	24 58 35.7	0.268
PHL 1049	0132+077	17.26	21.8±13	1 35 9.13	7 59 6.0	0.147
PHL 1127	0141+052	18.29	21.2±13	1 44 9.62	5 30 17.6	1.99
LB 9502	1510+237	18.90	20.6±18	15 12 37.43	23 34 1.8	1.887
LB 9308	0903+169	18.27	20.1±18	9 6 31.91	16 46 11.5	0.411
PHL 1186	0147+089	17.55	18.4±13	1 50 25.41	9 14 49.8	0.27
PHL 828	0044+030	16	17.9± 6	0 47 5.71	3 19 57.1	0.624
LB 8741	0847+190	16.6	17.9±18	8 50 29.41	18 53 49.2	0.568
LB 8948	0854+193	17.4	17.9±18	8 57 6.33	19 8 54.0	0.331
PHL 1092	0137+060	17	17.5±13	1 39 55.81	6 19 21.2	0.396
PHL 1027	0130+033	16.91	16.6±13	1 33 7.00	3 39 4.0	0.363
LB 8891	0852+181	18.21	15.8±18	8 55 37.87	17 55 27.0	1.013
LB 8863	0851+197	18	15.3±18	8 54 50.69	19 30 37.6	2.214
LB 9010	0856+186	18.3	13.0±18	8 58 50.48	18 26 4.6	1.711
PHL 3424	0131+055	18.25	12.0±18	1 33 44.45	5 47 53.8	1.847
RS 32	1336+264	18.91	12.0±22	13 38 23.13	26 13 52.4	0.341
LB 6158	0839+186	18.40?	11.3±18	8 42 49.26	18 30 17.7	2.052
RS 23	1333+286	18.74	9.2±22	13 36 13.33	28 24 59.0	1.908
PHL 3375	0128+074	18.00	5.8±18	1 31 2.38	7 43 40.7	0.390
LB 8775	0848+163	16.9	5.1±18	8 51 41.83	16 12 22.0	1.926
PHL 1222	0151+048	17.62	4.2±13	1 53 53.91	5 2 58.6	1.923
TON 621	1231+294	16	4.2±10	12 33 55.35	29 7 52.3	2.011
LB 19	1247+267	15.8	4.1±5	12 50 5.67	26 31 8.8	2.043
PHL 1226	0151+045	17.40	1.0±13	1 54 28.01	4 48 20.1	0.404

Figure 23: A table of proper motions for 40 quasars extracted by Varshni and Talbot from data compiled by W. J. Luyten (University of Minnesota, 1969) and included in the QSO catalogue by Hewitt and Burbidge (1993). (From the website www.LaserStars.org).

for the most part ignored in the literature. The fact of the matter is that proper motions have been seen, measured and catalogued—for *quasars!* It is neither new nor isolated.

This is a table drawn from data given in the 1993 quasar catalogue by Hewitt and Burbidge, giving proper motions of 40 quasars from highest downwards. This simply cannot be reconciled with the requirement that quasars lie at distances given by their redshifts, usually billions of light years away, unless we are to invoke implausibly high lateral speeds. Luyten's data also tellingly reveal preferred values in redshifts, which we discussed in chapter four. The first schedule of proper motions for quasars was compiled in 1969 by W.J. Luyten at the University of Minnesota, and his great work was sustained in the ensuing decades by a few resolute individuals, including, independently, Geoffrey Burbidge and Y.P. Varshni. Luyten was originally trying to determine the proximity of blue dwarf stars by deducing their proper motion, and later discovered that many of these "nearby stars" were in fact quasars. Not many people are granted telescope time to do this work, of course, because it is a persistent irritation to a Standard Model that demands quasars be at great cosmological remoteness, no matter what.

As a specific example, let's take the case of quasar TON 202, number two on the list. Not only has it the second fastest proper motion on the list (53 milliarcseconds per century), but also the highest apparent magnitude ($V = 15.63$). To put this in perspective, TON 202 moves at more than half the speed of Barnard's Star, the global record holder. If we look at these two numbers (high proper motion and apparent brightness) together, bearing in mind that Barnard's Star is less than 6 light years away from us, we are compelled to conclude that despite its high redshift, TON 202 is so close that it might even be within our galaxy! It is hardly a coincidence that the brightest quasar on the list is also very nearly the fastest.

Even more convincing are the data we acquire from a gamma ray burster (GRB) in the south-east corner of the local active galaxy M101. It has been positively identified as physically part of the galaxy, and as such, gives us food for thought. Distance to M101 is estimated with fair certainty to lie in the range 3.6 Mpc to 6.7 Mpc (Aside: Note the generous disparity in upper and lower limits). This makes it the closest AGN properly studied, and naturally enough also the richest field of observed quasar associations. A useful cross-correlation emerges from the GRB: It happens to be the brightest ever seen. The assertion of coincidence pales before these well-established associations.

The evidence for close proximity becomes even more compelling when other properties are considered, for example that TON 202 lies within 10° of the galactic plane (where we would expect to find members of the galaxy population), and that the proper motion is the same as that of neighbouring planetary nebula NGC 7293 (objects in the same region of the galaxy would move at the same speed). The most reasonable conclusion from these properties is that the quasar's redshift-given distance is spectacularly wrong.[*]

The illusion of remoteness can be further illustrated by examining the lateral motion of quasars projected to their redshift-given distances. The results are startling and must surely give grey hairs of those keenly promoting the authenticity of the Special Theory of Relativity. Y.P. Varshni of the University of Ottawa did the sums for three examples of quasars with well-established redshift values:

> Purely as an academic exercise, we calculate the transverse velocities required for the three quasars PHL 1033, LB 8956 and LB 8991 on the cosmological red shift hypothesis. We take the smallest value of proper motion within the uncertainty range and assume the Hubble Constant to be 50 km/s/Mpc and q_0 =0. Then we find that in terms of the velocity of light c, V_t = 760c, 5200c and 2300c for PHL 1033, LB 8956 and LB 8991 respectively. Needless to say these values are without physical significance and clearly indicate that the cosmological red shift hypothesis is completely untenable.

Just to clarify the point, three well-known quasars—by no means the highest redshift examples in our catalogues—are found to have lateral motions that are, respectively, 760 times, 5,200 times, and 2,300 *times the speed of light*. Holy smoke! Surely this alone renders the redshift-distance idea worthless? The final nail in the coffin of quasar redshift-distance (as if we needed another nail) is the measurement of the speed of jets. This too is proper motion, and is therefore a real effect. Over a period of 5 years from 1979 to 1984, John Biretta of the Space Telescope Science Institute measured the ejection of material from the quasar 3C 345. The increase in angular separation is directly observed and measured. To translate angular separation trigonometrically into linear distance, distance from Earth to 3C 345 is required.

Lest anyone starts wriggling uncomfortably now, let me reiterate that there is nothing peculiar about these measurements. These are first-order data, presented without modification; what you see is what you get. There can be no escaping the distress caused to physical science that results from applying redshift-distance to 3C 345. Then all hell breaks loose. It turns out that 3C

---

[*] See Halton Arp, *Catalogue of Discordant Redshift Associations* (Montreal: Apeiron, 2003) p 146.

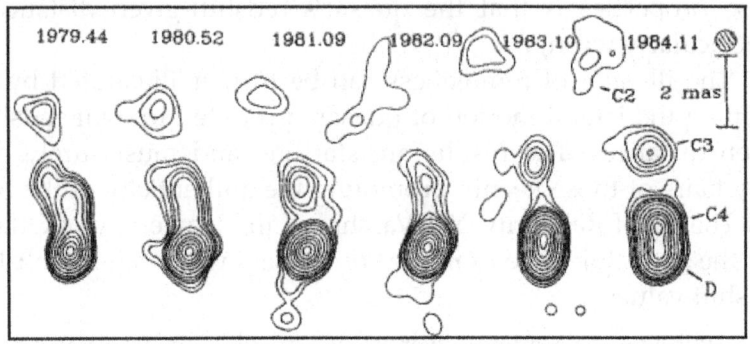

Figure 24: The angular separation from 1979 (left) through to 1984 (right) of host and ejected material in the quasar system 3C 345, measured by John Biretta. The direct linear rate of departure is a function of remoteness, by simple trigonometry. (Image courtesy of Dr Y P Varshni and John Biretta, STScI).

345 should be around five-and-a-half billion light years away, if redshift is to be believed. That number applied to the measured rate of angular departure means that the system is allegedly expanding *transversely* at a physical rate of seven times the speed of light. It matters not a whit whether we agree with the theoretically imposed absolute speed limit of c; the fact of the matter is that those who propose redshift-distance do so on the basis of Einstein's Relativity, and are therefore bound to bow before the absolute barrier imposed by light speed. They cannot have their cake and eat it. At redshift-given remoteness, 3C 345 surpasses the speed of light many times over. In terms of its own founding principles, the Hubble law is unambiguously falsified by Figure 24.

Once again I find myself in the strange and comfortless situation of having the need for any further argument rendered redundant. Redshift-based cosmological expansion is by this single, unassuming set of empirical measurements made utterly defunct. This is but one of several such examples in this book. Isn't it strange, therefore, that the ostriches of cosmology can continue to bury their heads in the sand of abject denial? In 1999, Halton Arp told the world,

> Of course, if one ignores contradictory observations, one can claim to have an 'elegant' or 'robust' theory. But it isn't science.

In 1956, Fritz Zwicky became the first astronomer to describe large-scale tidal effects characterising galaxies, in the form of "clouds, filaments, and jets of stars." He attributed these phenomena to ejection, caused by galaxy collisions. Renowned Armenian astronomer Viktor Ambartsumian tendered a very important alternative view, however, theorising the fissioning of celestial objects. According to Ambartsumian, galaxies split and multiply much in the manner of bacteria, although (thankfully!) at a far lower rate. This raised the possibility that galaxy-galaxy interactions and consequent tidal disturbances described by Zwicky, could well be caused primarily by the ejection of one object by another without their prior merging necessarily. [*]

I became interested in colliding galaxies some 15 years ago, but up until 5 years ago, I still believed them to be extremely rare occurrences. As a result of a fortuitous introduction to Iowa State University astrophysicist Curtis Struck, I learnt that I was very wrong about that. Collisions between galaxies are commonplace, with about 1 in 10 engaged in some stage of the collision process at any time. [†] More recently (2006), a team led by Francois Hammer used the multi-object GIRAFFE spectrograph at the Very Large Telescope in Chile to discover that fully 40% of distant galaxies were so disturbed that they are obviously collisional. Galaxy collisions are extremely commonplace. Caution is needed here. Chip told me in private correspondence that he had his "own theory of galaxy collisions." He didn't give details, but my guess is that he sees the galaxy transient event as a splitting up rather than a combining of galaxies. Either way, what we see in galaxies is that they are usually an interaction of more than one. The two spirals nearest to Earth, the Milky Way and Andromeda, both incorporate within their horizon of coherence several other smaller galaxian objects, and are of course themselves on a collision course with one another.

The principle to be borne in mind here is that galaxies in close proximity to one another interact vigorously, and there are some clues that hint at whether they converge or diverge. The clearest of these is the density of isophotes on a contour map of radiation intensity. The tendency would be for the object in focus

---

[*] Fritz Zwicky, *Ergeb. Exakten Naturwiss.* **29** (1956), 344)

[†] Curtis Struck, "Galaxy Collisions" *arxiv*: astro=ph/ 9908269. Struck published a revised and updated paper in 2005: "Galaxy Collisions – Dawn of a New Era" *arxiv*: astro-ph/0511335.

Figure 25: Seyfert's Sextet—four colliding galaxies and two by-standers. (Image courtesy of NASA, ESA, and STScI).

to be moving towards the cliff-face of bunched up isophote lines, and away from the plain where the lines are more spread out.[*] Another important point emerges from a study of colliding galaxies, or Transient Galaxy Dynamics if you want to be proper. We have already noted that every collision is evidence of at least two galaxies moving towards each other (militating against universal expansion at the scale of galaxies), and that collisions play a vital role in the morphology of galaxies. Collisions are hugely disruptive and leave nuclei of participants seething, very likely leading to the compaction of matter again into neutron stars at the core.

Curtis Struck states that "it is now thought that most galaxies experience several collisions or tidal interactions over the course of their lifetime that are strong enough to profoundly alter their structure and accelerate evolutionary processes." In addition, multiple interactions are now increasingly observed with up to 5 galaxies concurrently involved.

---

[*] Isophotes in astrophysics are analogous with *isobars* in meteorology, or *contour lines* on relief maps. They connect points of equal magnitude (brightness) on maps of astrophysical systems, and are invaluable in tracing physical connections between objects.

What we actually see in collisional systems is the central core of the developing super-galaxy becoming extremely agitated. An intriguing picture has emerged of the propagation of galaxies—they collide, become active, and eject quasars, which go on to become the nuclei of new galaxies. In an appealing fantasy, they mate, become pregnant, and give birth. Quite incredible!

So it would appear from well-documented observation that although galaxies can and do grow by aggregation, including collisions, they are born from high-redshift ejecta. Hubble Space Telescope observations further suggest that there is a hierarchical link between collisional systems and quasars. In the above paper, Struck says:

> HST observations...provide direct evidence that some, and the implication that most of the quasar hosts are collisional systems.

By the end of the millennium, major objections to collisions as fuelling mechanisms for quasars had been allayed, and transient galaxy dynamics emerged as a logical backdrop to the catalogued observations of Halton Arp.

"It Is No Mirage! ... Using ESO's Very Large Telescope and the W.M. Keck Observatory, astronomers... have discovered what appears to be the first known triplet of quasars." That's the excited wording of a European Southern Observatory press release announcing what appeared to be a new and important discovery in the cosmos. I have to admit, though, that I found it exciting for different reasons, because it is something that was predicted, subsequently observed, and catalogued consistently for decades by Halton Arp.

It was until this announcement denied equally consistently by the astrophysical mainstream. Team leader George Djorgovski of Caltech explains why the alignment is remarkable:

> Quasars are extremely rare objects. To find two of them so close together is very unlikely if they were randomly distributed in space. To find three is unprecedented.

Unprecedented, that is, unless you have looked at the photographs in Arp's Atlas of Peculiar Galaxies. Literally thousands of such arrangements have been observed and measured. The alignment of QSOs around an active galaxy is exactly what one would expect if they are indeed the ejected seeds of nascent galaxies, as Arp's images suggest.

The idea meets institutionalised resistance. Here is a quote from Jerry Jensen's 2004 paper "Supernovae Light Curves: An Argument for a New Distance Modulus" [*] (referring to Arp's abundant evidence of AGN—QSO links):

> That astronomers continue to doubt these convincing visual and tomographic radio images is one of the perplexing consequences of the power of preconceptions on the interpretive functions of the human mind.

In summary, we return to the doyen of reasonable science, Martín López-Corredoira. In his characteristic non-confrontational way, López-Corredoira summarises the unresolved issues that standard theory has with quasars in a 2009 paper entitled "Pending Problems in QSOs." [†] He puts the arguments into 10 categories. All except 2, 5, 9, and 10 have been discussed in this chapter and elsewhere in the book, so I shall mention just the salient points here:

1. Very high luminosity at high redshift. A single quasar would have to be as bright as hundreds of galaxy clusters together if their redshift really is cosmological in origin, placing them at extreme distances from Earth. Standard physics is unable to provide an explanation for hyperenergy, so it is left to hypothetical ultra-massive Black Holes (of around 10 billion solar masses) to fill the gap. However, such massively gravitating objects would produce relativistic acceleration of infalling matter, resulting necessarily in extreme, high-energy radiation signatures that are simply not observed. Moreover, the magnitude-redshift relation (basis of the Hubble law) is so widely dispersed that it shows no trend whatsoever and thus cannot be expressed as a Hubble diagram.

2. Host galaxies. The luminosities of quasar host galaxies, expected to be consistently high (brighter than normal galaxies), are by observation so erratic that stellar mass ranges from $M \sim 5 \times 10^{11}$ solar masses right down to zero (undetectable). The dynamical and chemical properties of

---

[*] Jerry W. Jensen, "Supernovae Light Curves: An Argument for a New Distance Modulus" *arxiv*: astro-ph/0404207.

[†] Martín López-Corredoira, "Pending Problems in QSOs" *arxiv*: astro-ph/0910.4297.

intra-galaxian gas are also anomalous for objects at such distance.

3. Age and metallicity of high redshift QSOs. The evolutionary phasing of some high-redshift quasars means they are older than the Universe. Accepting redshift distance brings severe problems for other benchmarks in the LCDMM, and there are a number of examples of redshift (or model) anomalies that remain to be solved.

4. Evolution or non-evolution of QSOs. The peculiar inverse relationship between redshift and luminosity of some quasars that are brighter at high z than others at low z cannot be reconciled with standard theory. Furthermore, the spread of luminosities cannot be reconciled with change in other properties, indicating that quasars do not evolve with redshift. "Big Bang requires that stars, QSOs, and galaxies in the early universe be 'primitive', meaning mostly metal-free, because it requires many generations of supernovae to build up metal content in stars. But the observations show the existence of even higher than solar metallicities in the 'earliest' QSOs and galaxies."

5. Triggering of activity. The conventional explanation of what triggers the interaction between quasars and their companion galaxies is variously ascribed to the formation of Black Holes and the accretion of gas upon their event horizons. However, the physics employed in Black Hole modelling are controversial to say the least, and certainly not well enough understood to justify anomalously distant energetic objects.

6. Superluminal motions. By simple trigonometry, the rate at which some high-redshift objects depart one another (proper motion and angular dispersion) exceeds the speed of light many times over. If Special Relativity is assumed, the conclusion is unavoidable: The redshift-distance is flawed. Theorists have attempted to accommodate these anomalies in the Standard Model, but their efforts are contrived and unconvincing. "There are some explanations. The so called relativistic beaming model (Rees 1967) assumes that there is one blob A which is fixed while blob B is traveling almost directly towards the observer with

speed V < c with an angle cos⁻¹(V/c) between the line of motion and the line B-observer. This leads to an apparent velocity of separation which may be greater than c. There is also another proposal in a gravitational bending scenario (Chitre & Narlikar 1979). However, both explanations share the common criticism of being contrived and having somewhat low probability ($\sim 10^{-4}$) (Narlikar & Chitre 1984). In the case of blazars, the superluminal motions in blazars can be statistically explained in the frame of the unification scheme of AGNs (Liu & Zhang 2007)."

7. Periodicity of redshifts. "In a homogeneous and isotropic Universe, we expect the redshift distribution of extragalactic objects to approximate a continuous and aperiodic distribution." However, this is not the case. When measured off the base of the host galaxy redshift, the redshifts of surrounding, nested quasars are measurably periodic.

8. Correlation with galaxies of lower redshift. The statistical prevalence of high-redshift quasars in near proximity to lower-redshift host galaxies is now well-established in observation, and strongly supported by contemporary high-volume surveys. Line-of-sight coincidence is both statistically and materially contra-indicated. In a number of classic examples, background signatures (absorption lines) for embedded quasars are ruled out and foreground signatures confirmed. The conclusive evidence is the spectral delineation of isophotes physically linking objects with widely varying redshifts.

9. Emission lines. The conventional interpretation is that quasars are merely very bright, hyperactive Seyfert galaxies (AGNs), with characteristic broad and narrow spectral emission lines. Whilst the standard Black Hole accretion disk scenario—where differences between type 1 and type 2 Seyferts are ascribed purely to the orientation of the torus—can be invoked to explain some of these phenomena, yet others remain anomalous to the model, and cannot be explained unless the model is complicated by ad hoc parameters.

10. Absorption lines. Once again, the Concordance Model proposes a solution that is only partially satisfactory. The

idea is that quasar absorption lines reflect the footprint of photons in the Inter Galactic Medium (IGM). Observations would tend to indicate otherwise. In particular, a comparison between the sight-line spectra of Gamma Ray Bursts (GRBs) and quasars with the same degree of redshift shows great disparity where one would expect them to be the same if they have crossed the same amount of space and traversed an equivalent amount of intergalactic media. In fact, the incidence is *four times* higher with GRBs! The matter is fairly technical and detailed exposition is beyond the scope of this work, so we shall leave it there with the suggestion that those with a deeper interest read López-Corredoira's paper and pursue the references therein.*

López-Corredoira concludes rather sombrely:

> Some cases which were claimed to be anomalous in the past have found an explanation in standard terms …There are, however, many papers in which no objections are found in the arguments and they present quite controversial objects, but due to the bad reputation of the topic, the community simply ignores them.

Don't worry, Martin, every dogma has its day!

There is something very important which should temper our analysis of quasars. *All of these quandaries about quasars were real only at their redshift-implied remoteness,* and would tend to disappear if the objects were in fact closer to our point of observation. It is only the pernicious assumptions of expansion and redshift-distance that lead to the energy dilemma infecting galaxian astrophysics. Remove those assumptions and the problems go magically away, but that of course means abandoning the model. Sadly, no one earning a living from that model, bar a few selfless heroes, is prepared to do so.

Notwithstanding all attempts to deny the astronomical community access to his work, Dr Arp managed to publish his catalogues, and they are at once excellent and damning of the Standard Model of Cosmology. It is sobering to reflect upon what the authorities in charge of the large American observatories were trying to achieve when they denied access to Halton Arp. They

---

* Martín López-Corredoira, "Pending Problems in QSOs" *arxiv*: astro-ph/0910.4297.

effectively stopped him from capturing images of *what is actually out there!* They are saying that there are some apparently bizarre astrophysical objects in the cosmos that the rest of us should not see or even be aware of. Please think about that carefully.

I beg the question: Should we astronomers ever become so servile that we mutely accept the imposition of scientific censorship?

> ***Discussion:*** *Quasars provide observational proof of intrinsic redshift and objectively falsify the Hubble law. Those favouring the Standard Model do not argue; they simply deny that the evidence exists, and decline to publish it.*

Chapter 6

The Microwave Background

Surround-sound radio

A uniform radio fog surrounding the Earth has been artificially impregnated with meaning, and consequently interpreted as an image of the primordial fireball. It is no such thing. "The statement that big bang theory explains the observed microwave background ... is to distort the meaning of words." (Geoffrey Burbidge).

◆ ◆ ◆ ◆ ◆ ◆ ❖ ◆ ◆ ◆ ◆ ◆ ◆

I was asked the question once, "How is it that all these astronomers—the majority—were persuaded that the so-called Cosmic Microwave Background Radiation (CMBR) came from the Big Bang?" The answer is plain—they didn't need to be persuaded, not in the least bit. Their model implicitly requires it, so come rain or shine, *cosmic* the microwaves will be! Clearly, there is a startling and very disturbing disparity between objectively described radiation images and those presented to us by the Standard Model; between the raw data acquired by an observatory's instruments and the buffed and polished, full colour pictures the rest of us get to see.

This chapter will be very nearly the longest in the book, for it is in the matter of the Cosmic Microwave Background Radiation that proprietors of the Standard Model crow their triumph the loudest, see the most all-embracing detail of the universe, and not coincidentally, practice the most passionate manipulation of observational evidence. This is a profoundly entrenched dogma, almost immune to logic and pure data, and it stretches me to my limits as a reasonable man.

I should not be so unbearably pompous as to proclaim my opposing views from the podium of ethics, nor should I ever poison my arguments with personal invective, but I'm only human. It's extremely difficult to avoid a troubling suspicion that so many wrong answers could not have been built into the model whilst under the relentless scrutiny of a regiment of supremely talented scientists, were they not predisposed to quite deliberately and purposefully achieve a desired objective *prior* to the evidence being produced. Cordiality and etiquette compel me to say that at the very least, it looks like wanton neglect. The people involved are without any doubt whatsoever completely sincere in what they do, and are unquestionably decent, ethical citizens of the world of science. Notwithstanding their propriety, we cannot deny that some data were manipulated and others suppressed, all with the tacit approval inherent in consensus. That is a huge problem for science, and I intend to expose it for what it is.

Briefly, it happened like this: Theorists had for years expected to find some sort of radiation picture of the primordial fireball, and to this end a high-powered team had assembled at Princeton University with the declared purpose of finding the heralded message from antiquity. Because it would have emanated from an unstructured, concentrated, hot gas, it was anticipated that the radiation would be evenly distributed across the sky. The Cosmic Background Radiation, hopefully, would present neither point of origin nor preferred direction, and furthermore, would be free of any links to foreground structures; if it is in actuality a true image of Big Bang, then north mirrors south, and east is indistinguishable from west. That is what they wished for. They wished also that when it finally was discovered, the rest of us would be so dazzled by the occasion that we too would declare that the surrounding long-wave radiation picture met all the requirements of their theory. Well some of us did not.

In 1965, the attention of Princeton cosmologist Robert Dicke was caught by the efforts of a pair of radio engineers at the Bell Laboratories in New Jersey. Arno Penzias and Bob Wilson were working on a telecommunications satellite project and were being bothered by persistent interference in the microwave band. No matter which way they pointed their oddly shaped radio antenna, there was an annoying hiss, and what was particularly frustrating to them was that it appeared to be coming from nowhere in particular. They saw no connection to anything useful or cosmologically significant in what was obviously radio noise, but when Bob Dicke heard about their serendipitous discovery, particularly that it was apparently isotropic, he and his team pounced.

History has in some ways been unfair to Arno Penzias and Bob Wilson, despite that they were in due course lavishly rewarded for their contribution. They were quietly and conscientiously doing their jobs when they accidentally tripped over something that turned out to be of immense potential value to the spin doctors of cosmology. The iniquity that followed was most certainly not their fault, and in truth, they played no part in it. They simply accepted praise and laurels for stumbling over a log in the woods, quite unaware at the time that it was a sought-after totem pole of tremendous worth to a particular culture. We can no more hold them accountable for the imposition of the Cosmic Microwave Background Radiation than we could blame Edwin Hubble for the Hubble law and the expanding universe model. They were merely unwitting pawns in a much bigger game.

Questions need to be asked. Does the presence of this fairly uniform radio noise not fit remarkably well with Big Bang predictions? Could anything else (besides a universal "explosion") have produced the observed effect? The answers that emerge from objective analysis are clearly "No, it does not" and "Yes it can."

Three key axioms are built into the theory (no, they are not predictions; BBT is not predictive): One, it should in all likelihood be a waveband longer than optical; two, it should be blackbody radiation;* and three, it should be isotropic, smoothly distributed across the sky. A fourth parameter came from calculations by Gamow, Alpher, and Herman: It was expected to have a temperature of between 28 and 50 kelvins.

We should add a fifth requirement overlooked by Big Bang theorists: Any candidate radiation under consideration for the

---

* Blackbody: An idealized, theoretical surface that absorbs and emits all radiation incident upon it, and has no capacity for reflection; stars are assumed to be blackbodies for purposes of describing stellar radiation. Blackbody radiation: Thermal radiation from a blackbody, which follows a characteristic curve of energy and temperature for any given wavelength; displays a Planck spectrum.

role of CMBR should show neither greater likelihood of having come from other, non-cosmological causes, nor present a better fit to the predictions of competing models. How does the Penzias and Wilson sample match up to these five requirements?

First, the good news: It is indeed in the microwave band, longer than red on the spectrum. Now for the bad news: It fails all the other requirements with degrees of severity varying from moderately strong to catastrophic. Any match with Big Bang Theory beyond the broad requirement of wavelength is entirely contrived, and should be rejected outright by sober scientists.

Let's face it; if the Big Bang chaps could show that the Cosmic Microwave Background Radiation really *is* a picture of the early Universe as it commenced expansion, they've got me! They will have established beyond doubt that the Universe is indeed non-static, and have given the Standard Model of Cosmology sorely needed "predictive" success. They would finally screw the lid down on the Big Bang critics that I seem to represent, and I shall have no choice but to knock my king over and concede the game. So why am I not worried? The answer is again quite simple. They are not even close, and it seems the culprit is once more that old bugbear, rash assumptions.

Geoffrey Burbidge calls a spade a spade. In 2008, he published a marvellous review of the development of modern cosmology entitled "A Realistic Cosmological Model Based on Observations and Some Theory Developed over the Last 90 Years." It is a timely reminder that Burbidge, one "B" of the all-time classic "B²FH" team that defined stellar nucleosynthesis in the 1950s, really is one of the greatest players in contemporary physics, and those that ignore him do so at their peril.

He gets right to it:

> From that time on one huge danger ... is that all of the work on the CMB has been carried on by observers who are absolutely convinced that whatever they find, they are quite sure where it came from. This has led to a bandwagon so overwhelming that alternative interpretations of the data are hardly ever mentioned, never taught, or discussed at meetings, or referred to in text books.[*]

---

[*] G. Burbidge, "A Realistic Cosmological Model Based on Observations and Some Theory Developed over the Last 90 Years" *arxiv*: astro-ph/0811.2402.

The fact of the matter is, if cosmologists were to limit their function to that of behaviourists examining the interaction of large objects (as I proposed in chapter one), none of the background radiation in any wavelength—there several such backgrounds—would have been labelled "cosmic" in the first place. The first images of the Cosmic Microwave Background Radiation were enthusiastically described by Cosmic Background Explorer (COBE) analyst George Smoot as being akin to the fingerprint of God, and in the words of Stephen Hawking, "the scientific discovery of the century, if not of all time."

I am fully aware that this is holy ground in cosmology, so I shall attempt to tread carefully. My approach will be to illustrate the same principle at work here that gave us the Hubble law, a phenomenon I have named Ideological Momentum:[*] From a theoretical model, preconception causes hasty adoption of fallacious data—in this case, the error-ridden results obtained by the pioneering COBE satellite observatory. Precisely as with redshift, contra evidence is openly and deliberately ignored, and enormous mathematical torque brought to bear to mould a fit with the preferred theoretical description of what things ought to look like. I shall attempt to show with valid scientific argument that the acclaimed microwave blackbody spectrum is neither predicted by prevailing theory nor measured by the COBE satellite. The finest scientists on Earth were apparently completely duped by unsubstantiated assumptions and broken data.

We shall before long be inundated with names, abbreviations and acronyms, so let's write a few down and familiarise ourselves. The Penzias and Wilson radio noise was soon grandiosely renamed the Cosmic Microwave Background Radiation or CMBR for short. Two important space missions were launched specifically to study the surrounding microwaves from above the atmosphere (thereby eliminating at least one layer of foreground "contamination"). These were the Cosmic Background Explorer (COBE), launched in November 1989, and the subsequent, much improved, Wilkinson Microwave Anisotropy Probe (WMAP) that took over from COBE in 2001.

---

[*] Ideological Momentum: The impetus of collective opinion; the tendency for supportive results to emerge from prior consensus or authority; also called 'the snowball effect'; a synthetic trend in which we impute meaning in things just because we *want* meaning to be there for whatever deeply held reason, and then take that meaning forward even when it has been objectively falsified.

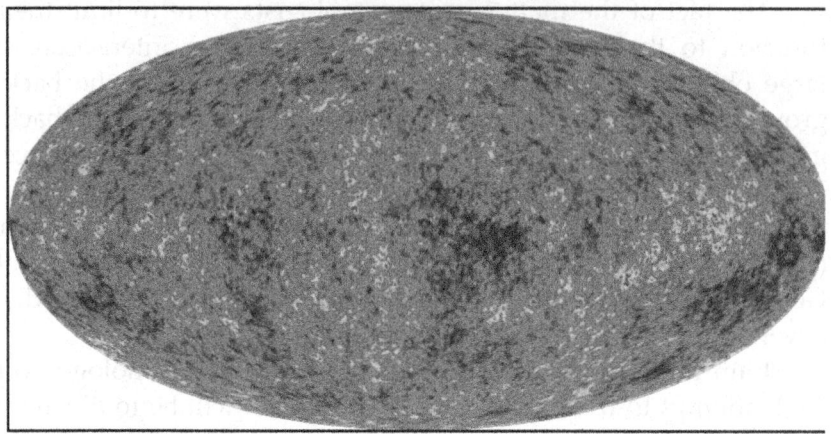

Figure 26: The CMBR as imaged by the WMAP satellite. It is obviously neither hemispherically symmetrical nor isotropic. Strangely, we see none of the expected shadows of intervening objects that should be clearly silhouetted if the radiation really is from the far background. (Wikimedia Commons public domain image courtesy of NASA).

You may well ask, given that WMAP operates at far higher resolution than COBE and actually superseded it, why I make such a fuss about what COBE discovered. Surely we can ignore the first mission and concentrate on WMAP? No we cannot. We'll miss some crucial issues if we do. Their instrument payloads are technically different, with different ways of reading the input signals, and it was COBE that established the broad principles while WMAP looks at the fine detail—crosses the "i"s and dots the "t"s, if you'll pardon my dislocated humour. It was from COBE measurements that John Mather drew his classic "perfect prediction" graph, and as in the case of cosmological redshift, an entire body of theory was verified by patently flawed data.

The whole matter of universal background radiation is exhausting, and certainly beyond the scope of this small work. I intend in this chapter to touch the nerve; I shall attempt to distil several essential points from the theoretical wash colouring cosmological radiation, and thereby expose the wishful science that frames it. We have circumextant radiation in all wavebands from radio to gamma, and most of it can be called isotropic to somebody or other's satisfaction. If this radiation is taken to be an image of something, or to contain an image of something, it is certainly not obvious. On first pass it is simply not possible to make out just what that subliminal message might be exactly. When the overall signal is converted into a power spectrum of hot and cold spots, it takes on the guise of what my grandmother, bless her, called "modern art," an abstraction in which only a specially trained eye would find meaning. Experience shows that such eyes are always around, eager to help.

The official COBE and WMAP teams in their turn ferociously interrogated the data coming in from the satellites. These microwave images are put through the wringer and creatively enhanced on an unprecedented scale. As the dust settles on each annual release of data, volumes of analyses are written up and published. This runs to many thousands of published papers— ask me, I've been engaged in a statistical analysis of WMAP results for several years now and I am quite flabbergasted by the sheer volume of work being produced. It's quite amazing what they've found, and incredible *how* they found it. But the fact of the matter is that all these conclusions could have been reached without there necessarily being any link to the Big Bang, and the way is kept clear by ignoring contrary findings. Many diligent interpretations are put aside as anomalies, to be dealt with at some unspecified time in the future.

Assuming that there *is* primeval radiation coming our way at a temperature of 3K, how would we recognise it? By definition it would have to originate from far behind the galaxies and nebulae that surround us, and would at best be mixed up with radiation from those things. I suppose the most obvious clue would be that foreground objects would be backlit, casting shadows on the sky. We do not see any shadows!

The Universe surrounds us; there are stars everywhere. Surely all that light counts for something, even if it's not visible? The radiation that so perplexed Heinrich Olbers has been reliably estimated to have a blackbody spectrum at 3K; moreover, matter beyond the Kuiper Belt (which encloses the Solar System) would, upon being heated by the Sun, radiate back to us at a temperature of about 3 kelvins. Consider also the rather astounding proposal made by Paul Marmet and others that the so-called "missing mass" saturating the universe radiates at energy of 3K (and there's good reason to believe that it would).[*] If that's the case, then Dark Matter is not invisible, non-baryonic matter after all. It's actually normal, radiant matter, and we have been staring at a microwave picture of it all along! It's certainly more plausible than the absurdities surrounding non-baryonic Dark Matter. The fact remains that there's an enveloping fog of 3K radiation coming at us from all sides, and there's nothing that links any of it to a Big Bang other than a temperature that's about a seventeenth of what Gamow's team predicted.

Andre Assis and Marcos Neves published an excellent summary of background radiation temperatures in the journal Apeiron in 1995, and I am indebted to them for great insight into the folly of Peebles and Dicke.[†] Well prior to that, the equilibrium temperature of interstellar space was calculated to be around 3K and predicted independently by Guillaume, Eddington, Regener, McKellar, Nernst, Herzberg, Finlay-Freundlich, and others. Even arch-Relativist Arthur Eddington made a strong case for a static Universe, before he too was converted to the fashionable new faith. He meticulously calculated what radiation we should find imprinted in interstellar space, and why it should be so. He found that ambient starlight would settle to an equilibrium temperature of 3.2 Kelvins, without the need for a fireball or subsequent expansion. This was recalculated by Regener in 1933 and refined to a temperature of 2.8K, still without the need for Big Bang or expansion. In a remarkable paper published in 1937, Walther Nernst made the following statement:

> In Regener's important work cited above it is found the fact that, in the universe, a celestial body that absorbs cosmic radiation must be heated to 2.8°K.[‡]

Clearly, the background radio signal eventually discovered by Penzias and Wilson had a far better fit to static Universe models heated by natural objects than it did the expanding fireball

---

[*] See Olbers' paradox in the glossary, and A.K.T. Assis, "On Hubble's Law of Redshift, Olbers' Paradox, and the Cosmic Background Radiation" *Apeiron* No. **12**, Winter 1992.

[†] See Andre Assis and Marcos Neves, "History of 2.7K Temperature Prior to Penzias and Wilson" *Apeiron* Vol. 2 Nr. 3 (July 1995), pp. 79-84.

[‡] W. Nernst, "Weitere prüfung der annahme lines stationären zustandes im weltall," *Zeitschrift für Physik* **106** (1937), 633-661 (translated from the original German by Peter Huber *et al.*).

model of George Gamow. In 1961, Gamow published his calculation of the background temperature: 50 kelvins, more than 17 times greater than that seen by Penzias and Wilson. It was undoubtedly Gamow's legendary charisma and stage persona that carried the day for the extraordinary version of events that would triumph in cosmology. It was most certainly not because the facts were on his side.

The microwave background is out there, of that there is no doubt. Anyone with a radio receiver tuned to the right frequency can hear it, and it bothers us visibly as "snow" on our television screens, so we have no need to further verify its existence. But, before we rush off and label it relic radiation from the Big Bang, there are two things that we need to do with CMBR, and they are firstly, establish what it can tell us, and secondly, guess where it came from.

To deal with the first task, we have to find out whether there is lumpiness in the radiation; very smooth waves are going to tell us very little. The first step would be to scan the background for variations or concentrations of temperature. Anisotropy will be reflected in fluctuating levels of energy. We should look carefully for any sign of asymmetry in the distribution—does left mirror right, and top match bottom?

Both COBE and WMAP have established dipole anisotropy in the microwave background indicating a motion in the direction of the Virgo supercluster. They have also revealed a slight general anisotropy in the radiation, a gentle lumpiness that appeals to the contention that it originated in the almost smooth plasma soup in the immediate aftermath of the Big Bang. Fair enough.

But the unevenness of the radiation reflects also the natural anisotropy of the background universe, a remarkable rendition of sharp points like planets, stars, nebulae, and galaxies against the vast interstellar voids. It is just what one would expect of radiation permeating the cosmos around us.

I think Martín López-Corredoira explains this issue really well:

> Regarding CMBR anisotropies, the power spectrum is just a curve with two or three clear peaks that could be parameterised with (about) 10 parameters (three parameters per peak: central position, width, height). If we allow certain range or errors (each peak has important relative error bars, which are very large in the 2nd, 3rd and beyond—

indeed, after the 3rd peak the noise dominates), it is possible to parameterise a curve like this with somewhat fewer parameters within the errors. Standard concordance cosmology reproduces the curve with six parameters (there are indeed about 20 parameters; but the most important ones are six in number; the rest of them produce small dependence), with some problems to reproduce the very large scale fluctuations. Nonetheless, there are also other papers which reproduce the same WMAP data with totally different cosmologies with a similar number of free parameters: *e.g.*, Narlikar *et al.* (2003), McGaugh (2004). The fact that different cosmologies with different elements can fit the same data (with a similar number of free parameters to fit) indicates that the number of independent numbers in the information provided by WMAP data is comparable to the number of free parameters in any of the theories.[*]

This list could go on and on; the bottom line is that if we graciously allow that the CMBR makes at least *some* sense — invisible "dark sense" perhaps? — then we must concede that it makes much more sense as the limiting temperature of space heated by ambient starlight and radiation from astrophysical structures, including even the Earth itself, than the signature of a hypothesised primordial explosion.

The Wilkinson Microwave Anisotropy Probe, in its very name, carries the implicit admission that focus of interest has shifted dramatically, from the originally much-vaunted isotropy in the image, to the complete opposite — anisotropy, or asymmetrical patterns, in the microwave image. Although the background radiation was defined by its isotropy, *all* analysis is now concerned with details that prove that it is actually *not* isotropic![†]

Please allow me a diversion here. One of the most profoundly important papers published on the matter of the CMBR and what the anisotropies might actually mean arrived innocently enough on the desks of weary astronomers in 2007. It is unique in one sense — it makes an extremely pertinent point without resorting to a single properly scientific argument. It emerged in the wake of a conference on outstanding issues in cosmology held at Imperial College, London, ironically the aca-

---

[*] Martín López-Corredoira, "The Sociology of Modern Cosmology" *arxiv*: astro-ph/0812.0537.

[†] If need be, please refer to the glossary at the back of this book for an explanation of these terms.

demic home of rebel theoretician and author of Faster than the Speed of Light, Joao Magueijo. It may have a ghostly lead author from an untraceable cyber university, but this paper is the most truthful account of CMBR analysis I've ever read. It really is worth absorbing, for beneath the wry humour lies subliminal argument with gritty substance. Here is part of the conclusion:

> Simply extending our analysis, we suggest that cross correlations between CMB and any other map of a Solar System body, image of a person, or an image of an animal will be detected at some statistical significance. Finally, we wish to comment on an existing suggestion in the literature that there are hidden messages from the creator in the WMAP data since it can be thought of as a billboard visible throughout the Universe so a message is likely to be encoded within the intensity fluctuations. In the extreme case that this message is in fact an image of the creator herself hidden within the cosmic noise, we suggest that it may be possible to establish this image through a large number of cross-correlations of input images (of people, animals, spirits, or combinations) that are used in a likelihood analysis. The image, or an image reconstructed from an ensemble, that maximizes the likelihood in such a comparison can be considered to be the best reconstructed image of the creator. In fact, we would not be surprised if a spherically projected image of a famous celebrity correlates with WMAP at some high significance. Thus, it is left to the reader to establish using a complicated theoretical argument if that celebrity is the creator whose image is then hidden in WMAP.[*]

Many a true word spoken in jest…

Back to business. The continuous stream of anomalous results from WMAP data is invariably treated in one of two ways: Either it is ignored, or the theoretical basis of the prediction is revised to give the desired effect, thereby changing it from prediction to arbitrary parameter. For example, the model predicts a definite Gaussian[†] distribution in the power spectrum. Observation clearly contradicts it:

---

[*] Check it out: "On the origin of the cosmic microwave background anisotropies" attributed to a team led by Ria Follop, Institute of Fundamental and Outstanding Questions, Department of Cosmology and Metaphysics, Online University, Internet. (*arxiv*: astro-ph/0703806).

[†] Gaussian distribution: Essentially, a normal distribution of data points on a graph, with scatter symmetrical across the trend line.

> We present measurements of the clustering of hot and cold patches in the microwave background sky as measured from the Wilkinson Microwave Anisotropy Probe (WMAP) five-year data. These measurements are compared with theoretical predictions which assume that the cosmological signal obeys Gaussian statistics. We find significant differences from the simplest Gaussian-based prediction.[*]

Background radiation comes in all wavebands of light, and most are considered to be ambient starlight of one kind or another. The single exception appears to be the CMBR, which is held by consensus to be a unique picture of the primordial fireball, coming from behind all observed astrophysical structure. Here, a study stops just short of the microwave band in declaring that,

> ...the extragalactic background light (EBL) from the far infrared through the visible and extending into the ultraviolet is thought to be dominated by starlight, either through direct emission or through absorption and re-radiation by dust.

If all other frequencies of background emission are attributed to some form of starlight, why should microwaves be exclusively non-stellar?[†]

The results of WMAP data are equivocal. As a child I would lie on my back and look at the gathering cumulous progenitors of an African summer storm, and wonder whether the evolving familiar shapes of artificial objects that I saw there had some metaphysical, paranormal connection with their equivalents in my daily life. How did 3-D reflections of jet planes and prime ministers take their reference and mould the clouds? I'm a grown-up now, and I'm still wondering the same thing, only this time it's in the far more serious theatre of microwave radiation analysis. Is there any link between the shapes the WMAP team sees and reality? There must be, but unfortunately for them, it's far less spectacular and not nearly as remote as they would like it to be. However, that will no doubt come out in the wash, sooner or later. Their first line of duty was to the prevailing cosmological inter-

---

[*] Graziano Rossi *et al.*, "Non-Gaussian Distribution and Clustering of Hot and Cold Pixels in the WMAP Five-Year Sky" *arxiv*: astro-ph/0906.2190.

[†] Justin D. Finke, Soebur Razzaque, Charles D. Dermer, "Modeling the Extragalactic Background Light from Stars and Dust" *arxiv*: astro-ph/0905.1115.

pretations of the Universe. Do the data fit with an accepted model?

On the face of it, the initial impressions from both COBE and WMAP did not support the pre-supposed Big Bang Theory. What was wrong? To be perfectly frank, nearly everything was wrong. Firstly, the equilibrium energy discovered in 1965 was by a factor of many thousands at variance with the predictions published by Gamow in 1961. Secondly, it was not in fact isotropic, but showed anisotropies aligned with local astrophysical structures. Despite the fact that the CMBR is supposed to be isotropic, and was hailed as the missing link because of its perceived isotropy, all current analysis is concerned with *anisotropies* in the microwave picture. An enormous amount of effort is being expended on studying properties of an image that, by definition, should not be there.[*]

Thirdly, both sets of results show a preferred direction towards Virgo at a gross velocity of 600 km per second, famously named the "Axis of Evil" by theoretical physicists Joao Magueijo and Kate Land. Preferred direction is expressly forbidden by Big Bang Theory. In their paper "The Axis of Evil revisited," Land and Magueijo state in the abstract:

> We find strong (and sometimes decisive) evidence for the 'Axis of Evil' in almost all renditions of the WMAP data.[†]

Fourthly, there is a complete absence of gravitational lensing effects in the distribution of cool spots in the radiation. We should also note that mooted Sunyaev-Zeldovitch effects fall far short of the model's predictions.

Before we go any further, let us deal with the question of blackbody. It is claimed that the BBT predicts a perfect blackbody for the radiation. It does not. There are two reasons why I make this claim. Firstly, Big Bang Theory cannot be described as predictive. Secondly, the early Universe described in the LCDM model would in terms of current understanding of thermodynamics, *not* have produced a blackbody spectrum. Please allow me a small

---

[*] Stefan-Boltzmann's law has the form $M_e = \sigma T^4$ where M is the emitted radiant flux density and T is temperature. Thus, energy has a fourth-power relationship with temperature, making the disparity between observation and prediction far worse.

[†] Kate Land and Joao Magueijo, "The Axis of Evil revisited" *arxiv*: astro-ph/0611518.

digression to look at this more closely, because these two points alone effectively remove any viable connection between the CMBR and Big Bang.

On the first count, I refer once again to someone who has "been there, done that," Geoffrey Burbidge. In his 2008 paper "A Realistic Cosmological Model Based on Observations and Some Theory Developed over the Last 90 Years" Burbidge reminds us that it is all too easy to mistake axioms for predictions:

> The statement that big bang theory explains the observed microwave background ... is to distort the meaning of words. Explanations in science are normally considered like theorems in mathematics, to flow deductively from axioms and not to be restatements of axioms themselves. Thus the radiation-dominated early universe is an axiom of big bang cosmology, and the supposed explanation of the CMB ... is a restatement of that axiom.[*]

Mike Disney, professor of astronomy at Cardiff University, argues the same point from a different angle.

> It is true that the modern subject has taken a turn for the better, if only because observers can now build instruments to deliberately test out cosmological ideas where, in the past, cosmologists could only wait for serendipitous evidence to turn up. On the other hand, in order to explain some of the surprising new observations, theoreticians have had to conjure up heroic and insubstantial notions such as 'Dark Matter' and 'Dark Energy' which supposedly overwhelm, by a hundred to one, the 'ordinary' universe we can actually detect. Interested laymen are bound to ask whether we should be more impressed by the new observations, or more dismayed by the theoretical djinns which have been conjured up to explain them.

In essence, what Disney is saying is that as long as free parameters exceed actual measurements—and they do so substantially in BBT—then any postulate created by twiddling those parameters cannot sensibly be called a prediction. They are simply axioms of the theory devised by manipulating the primary assumptions in a completely *ad hoc* manner until the twiddlers are satisfied.

On the second issue, we should remember that all starlight is taken in astrophysics to be blackbody radiation. There is nothing

---

[*] G. Burbidge, "A Realistic Cosmological Model Based on Observations and Some Theory Developed over the Last 90 Years" *arxiv*: astro-ph/0811.2402.

unique about the blackbody nature of the CMBR. It merely indicates that it is in a state of thermal equilibrium; it radiates everything it absorbs. Notwithstanding that, the astonishing fact is that the primordial event would in any case most certainly *not* have produced blackbody radiation. It falls foul of the laws of thermodynamics, and that is something even the most foolhardy cosmologist should think twice about. This revelation was delivered to a stunned audience at the 2nd Crisis in Cosmology Conference (CCC2) in 2008, in a paper entitled "The Cosmic Microwave Background Radiation does NOT prove that the Hot Big Bang Theory is correct." The conclusions the author reaches are of such immense importance to cosmology that I shall spend some time with them, and quote rather more than usual from his paper.

Bernard Bligh is a specialist in thermodynamics and nuclear physics, and over a period of years, he applied his mind to the question of relic radiation. In the abstract of that paper, Bligh introduces his arguments:

> It has frequently been asserted that the discovery of the Cosmic Microwave Background Radiation (CMB) by Penzias and Wilson is proof of the validity of the Hot Big Bang Theory of the origin of the Universe. In reality this is not the case because the expansion of the Universe at the time of the supposed 'Fireball' would not produce the perfect black-body radiation which is actually observed. This problem with the CMB has been pointed out before by Mitchell (1994) but the present study establishes the argument by means of rigorous thermodynamic calculations.[*]

Bligh thoroughly states the principles upon which his arguments are based, and using established physics, traces the postulated thermodynamic path from theoretical decoupling of the primordial radiation through to the present day. He states:

> It is deduced that the resulting radiation would not have a black body spectrum, but a smeared spectrum ... The substance of this paper is to give the objection a firm thermodynamic basis. I put it to my critics that this argument is unassailable ... it is based on good sound science. Therefore if a theory is postulated which contravenes the Laws of Thermodynamics, cosmologists cannot wriggle out of

---

[*] Bernard R. Bligh, "The Cosmic Microwave Background Radiation does NOT prove that the Hot Big Bang Theory is correct" accepted for Franklin Potter, editor, *Proceedings of the 2nd Crisis in Cosmology Conference*, Astronomical Society of the Pacific Conference Proceedings, 2009.

this fallacy by pleading that Thermodynamics is not relevant to the theory.

He goes on:

> The CMB is said to have been produced at the time of 'decoupling' when the electron density in the primeval Universe was very small. The radiation generated at that epoch would have had a black-body spectrum. Three cases are analysed when the electron density approached zero; ... Wien's law is applied to calculate the fall in temperature of the radiation for each case, assuming that the black-body spectrum is maintained. According to the Hot Big Bang Theory the three cases should all arrive at 2.72 K, but they do not. The conclusion is that the CMB spectrum ought to be 'smeared' and not the almost perfect blackbody curve, which is actually observed. Therefore the Hot Big Bang Theory fails this test.

If Bernard Bligh is right—and no one has yet managed to find chinks in his arguments—then you will agree that this paper is dynamite under the bastion of universal expansion. Committed empiricist Pierre-Marie Robitaille takes an altogether more fundamental approach:

> Kirchhoff's law was so powerful that it would become the foundation of contemporary astrophysics. By applying this formulation, the surface temperatures of all the stars could be evaluated, with the same ease as measuring the temperature of a brick-lined oven. For astrophysics, this meant that any object could produce a blackbody spectrum. All that was required was mathematics and the invocation of thermal equilibrium...[*]

The much-vaunted "perfect fit" curve published by Mather *et al.* in 1991 is indeed a wonderfully precise match, the result of years of intense scrutiny. In private correspondence, my friend and helmsman Paul Jackson shared the experience.

> I remember following with excitement the build-up to that figure, which happened over many years from 1967. Wobbly non-satellite data plots first emerged with huge error bars. The eventual super-accurate FIRAS results of COBE had error bars so small that they had to be multiplied by 400 to be visible on the plot.

---

[*] Pierre-Marie Robitaille, "COBE: A Radiological Analysis" *Progress in Physics*, Vol. **4**, (October 2009), pp. 17-42.

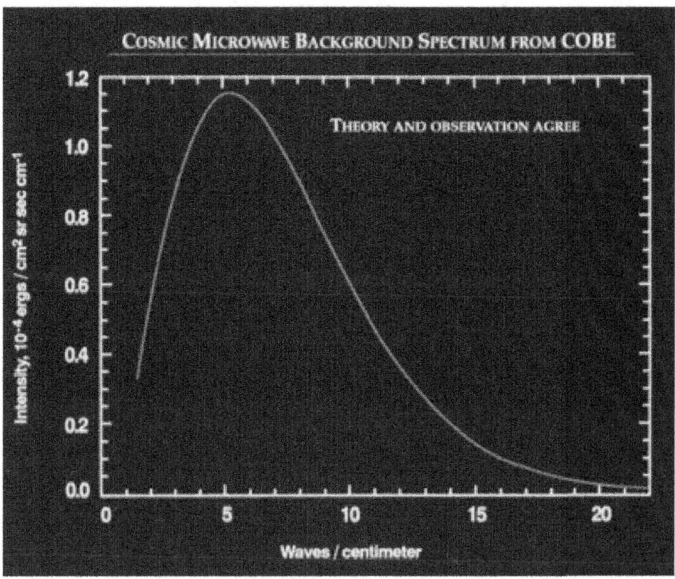

Figure 27: The much-vaunted "prediction-matches-observation" plot made famous by John Mather. Both the prediction and the observation are controversial. (Wikimedia Commons public domain diagram by kind courtesy of NASA).

The plot compares the model's requirement of a precise blackbody spectrum with COBE's mooted measurements of a blackbody in the microwaves. The questions seldom asked are "how accurate were the COBE readings?" and "is a blackbody in fact predicted?" If we can answer either in the negative, the theory fails; if both are negative the whole matter becomes a farce.

COBE carried three instruments. One of them, the Far Infrared Absolute Spectrometer (FIRAS), was designed to do nought but measure temperature and compare the radiation spectrum with that of a precise blackbody. It was from COBE's FIRAS measurements that Mather drew his plot.

Sometimes it's pure happenstance that grants one insight to workings of another man's craft. Perhaps it's a function of being in the right place at the right time. I exist because I happen to have crawled out of my cocoon at precisely the right moment in the Earth's history for me to flourish. The 10,000-year fragment of history that brewed up civilisation and technology represents 0.00022% of the age of the Earth. I landed in it.

Shortly before he died in 2009, my academic advisor Tony Bray nonchalantly changed the whole complexion of my understanding when he said, "Here, take a look at this." In his hand was a print of a paper by specialist professor of radiology, Pierre-Marie Robitaille. It's called "COBE: A Radiological Analysis." It turns out that this paper is a vital link in the chain of argument surrounding the way that the microwave radiation is measured, and equally importantly, how patterns are extracted from the fuzz.

The diagram (Figure 27) shows the incredibly precise blackbody spectrum supposedly measured by COBE's FIRAS instrument (plotted as intensity against frequency). However, that's all it does—it traces a perfect blackbody curve—and as we shall soon see, that in itself does not say anything meaningful about BBT or expansion. Robitaille, an acknowledged and acclaimed authority in the field of radiation measurement, gives an intensive appraisal of the design failures of that particular instrument, leading to the shocking revelation that methods employed by the COBE team were so flawed as to be utterly misleading.

Please allow me to quote rather extensively from this paper (with sincere apologies for the technical complexity). I would ask that you stay with me here, please. This is where my drill hits the nerve. Firstly, Robitaille deals with the inadequacies of the FIRAS instrument itself (I added the emphasis):[*]

> Data released from FIRAS has been met with nearly universal admiration. However, a thorough review of the literature reveals significant problems with this instrument. FIRAS was designed to function as a differential radiometer, wherein the sky signal could be nulled by the reference horn, Ical. The null point occurred at an Ical temperature of 2.759 K. This was 34 mK above the reported sky temperature, 2.725 +/– 0.001 K, a value where the null should ideally have formed. In addition, an 18 mK error existed between the thermometers in Ical, along with a drift in temperature of ~3 mK. A 5 mK error could be attributed to Xcal; while a 4 mK error was found in the frequency scale. A direct treatment of all these systematic errors would lead to a ~64 mK error bar in the microwave background temperature. The FIRAS team reported ~1 mK, despite the presence of such systematic errors. But a 1 mK error does

---

[*] Pierre-Marie Robitaille, "COBE: A Radiological Analysis" *Progress in Physics* **4** (October 2009), pp. 17-42.

not properly reflect the experimental state of this spectro-photometer. In the end, all errors were essentially trans-ferred into the calibration files, giving the appearance of better performance than actually obtained [...] Neglecting to fully evaluate FIRAS prior to the mission, the FIRAS team attempts to do so, on the ground, in highly limited fashion, with a duplicate Xcal, nearly 10 years after launch ... *Despite popular belief to the contrary, COBE has not proven that the microwave background originates from the universe and represents the remnants of creation.*

Robitaille goes into considerable and intimate detail on as-pects of the FIRAS instrument which render its measurements fraught with inaccuracies, but it would be sensible to keep this précis fairly brief. Next, he makes a revealing summary of the blackbody issue:

One hundred and fifty years have now passed, since Kirchhoff first advanced the law upon which the validity of the microwave background temperature rests. His law of thermal emission stated that radiation, at equilibrium with the walls of an enclosure, was always black, or nor-mal. This was true in a manner independent of the nature of the enclosure. Kirchhoff's law was so powerful that it would become the foundation of contemporary astrophys-ics. By applying this formulation, the surface temperatures of all the stars could be evaluated, with the same ease as measuring the temperature of a brick-lined oven... How-ever, since blackbody radiation only required enclosure and was independent of the nature of the walls, Planck did not link this process to a specific physical cause. For astro-physics, this meant that any object could produce a black-body spectrum. All that was required was mathematics and the invocation of thermal equilibrium. Even the re-quirement for enclosure was soon discarded. Processes oc-curring far out of equilibrium, such as the radiation of a star, and the alleged expansion of the universe, were thought to be suitable candidates for the application of the laws of thermal emission. To aggravate the situation, Kirchhoff had erred in his claim of universality. In actual-ity, blackbody radiation was not universal. It was limited to an idealized case which, at the time, was best repre-sented by graphite, soot, or carbon black. Nothing on Earth has been able to generate the elusive blackbody over the entire frequency range and for all temperatures.

This casts further material doubt upon the validity of the blackbody prediction. Right now, we are naturally more anxious to hear about the invisible error bars. Was the match between theory and observation really that marvellously precise?

> Despite the presence of systematic errors, the FIRAS team is able to essentially sidestep the recordings of their thermometers and overcome their inaccuracy. [...] The FIRAS team presents a dozen values for the microwave background temperature, using varying methods. This occurs over a span of 13 years. Each time, there is a striking recalculation of error bars. In the end, the final error on the microwave background temperature drops by nearly two orders of magnitude from 60 mK to 0.65 mK. Yet, as will be seen below ... FIRAS was unable to yield proper nulls ... Despite the subsequent existence of systematic errors, the FIRAS team minimizes error bars [...] Relative to error bars, the result obtained, using an average of many methods was analogous to ignoring the existence of known temperature error in the reference calibrators Xcal and Ical. The existence of imperfect nulls was also dismissed, as were all interferograms obtained while the Earth was directly illuminating FIRAS... It is well established, not only in physics, but across the sciences, that systematic errors can be extremely difficult, even impossible, to detect. Consequently, one must not dismiss those systematic errors which are evident ... This treatment would discount attempts to lower the error bar to 1 mK in the final FIRAS report.

The stage is now set for the *coup de grace*, an exposé of the precarious vulnerability of the famous Mather plot when its credentials are examined. After listing 15 serious problems with the COBE instrumentation, Robitaille brings it into focus in a single devastating sentence (my emphasis):

> Given the systematic errors on Xcal, Ical, the frequency drift, and the null temperature, it is reasonable to ascertain that the FIRAS microwave background temperature has a significant error bar. As such, an error on the order of 64 mK represents a best case scenario, especially in light of the dismissal/lack of data at low frequency. The report of a microwave temperature of 2.725 ± 0.001 K does not accurately reflect the extent of the problems with the FIRAS instrument. Furthermore, the absolute temperature of the microwave background will end up being higher than 2.725 K, when measured without the effect of diffraction, and when data below 2 cm^{-1} is included. *Contrary to popular belief, the FIRAS instrument did not record the most perfect blackbody spectrum in the history of science.*

After so thoroughly and meticulously demolishing the COBE data and the vacuous ideology built upon it, Robitaille's concluding statements are muted, almost demure. It's seems to me he's grown tired of shouting over the babble.

> Relative to the DMR (Differential Microwave Radiometers), the problems mirror, to a large extent, those I voiced earlier with WMAP. The most pressing questions are centred on the ability to remove the quadrupole from the maps of the sky. In so doing, it is clear that a systematic residual will be created, which can easily be confounded for true multipoles. In the end, the methods to process the anisotropy maps are likely to be 'creating anisotropy' where none previously existed.[*]

In performing a statistical review of 1,000 published WMAP papers, I found that nearly three quarters of the works in my sample use the words "anomaly," "anomalous," or simile in the abstract or introduction. Over tea one day, research supervisor Tony Bray and I were discussing the discordant results emerging from WMAP analysis, and I was trying to figure out how I could tabulate all these problems. He made a brilliant suggestion: Instead of trying to reinvent the wheel, why not let the analysts themselves do the work? All I need do is identify some key words indicating that the authors were puzzled, and apply a statistical filter. The word "anomaly" immediately sprang to mind, but even I was surprised to find just how many WMAP teamsters were not getting the answers they wanted. This is a worrying indication that most of the mathematical effort that follows is strapped onto a massage table, trying to get the data to fit preconceptions.

What should we read into this? Data themselves cannot be intrinsically anomalous—they are simply the results of measurement. An anomaly emerges *only when data are compared with some pre-conceived theoretical benchmark*, in other words, do the measurements meet expectations? If they do so consistently, that may be sufficient to declare the underlying hypothesis a law. If they do *not* meet expectations however, one would describe them as *anomalous* and should by rights then set about modifying or abandoning the observationally falsified theory.

---

[*] The foregoing verbatim quotes are taken courtesy of Pierre-Marie Robitaille from his paper, "COBE: A Radiological Analysis" *Progress in Physics* **4** (October 2009), pp. 17-42.

Before we take this any further, we should discuss a very useful device employed by engineers—signal-to-noise ratio (SNR). In any data transmission, there is a degree of unwanted pollution in the received image, for example static interference in radio messages. Much of the effort going into analysing the CMBR data is centred on stripping the "signal" away from the "noise."

How would we know that the microwave background is cosmological? What is it that indicates to us with certainty that the microwave radiation surrounding us came from the Big Bang? If we have the means (or *a priori* information) to determine the specifications of the pristine signal at source, we can identify what in the potpourri of incoming data might be the intended "signal," and what might comprise "noise" from environmental influences or instrument effects. Our intention as engineers should be to optimise the signal-to-noise ratio, in other words, to get the maximum (desired) signal for the least possible amount of (unwanted) noise. However, we faced an insurmountable problem: We have no idea beyond conjecture what the signal would have been at source.

You will recall that radio engineers Penzias and Wilson, who subsequently achieved undying renown as discoverers of the Cosmic Microwave Background Radiation, were at first annoyed by it. They saw it (quite correctly) as unwanted static interference in their satellite communications signals, and tried their best to factor it out of their data. In other words, they knew what the signal should be, because they had it at source, and they were consequently able to define the entire microwave signal as noise.

What changed? Why and how did static interference miraculously metamorphose into the Picture of Creation? How on Earth did undiluted *noise* in one instant become the heralded *signal* in the next? The signal itself did not change; the frame of expectations did. The microwave picture was more or less evenly spread over the sky, and really, that was all that was needed. The rest would come on the massage table.

Now that we have that safely out of the way, let us return to our signal-to-noise ratios. On the assumption that the signal came from behind everything we can see, the initial and ongoing efforts have focussed on removing what is termed "foreground contamination." Nothing was spared. That, my friends, was when they threw the baby out and kept the most unsavoury part of the bathwater. A layman's definition of signal-to-noise ratio might be "the proportion of good stuff (signal) that comes with the static (noise)," or perhaps, "the ratio of desired signal to undesired signal." The "desire" for "good stuff" is not hard science; it's an emotionally top-heavy, psychological need. Optimising signal-to-noise ratio of the microwave background image is entirely subjective, depending on arbitrarily defined signal parameters, and thereafter by elimination, the noise parameters.

In New Scientist of 30 April 2005, tucked away between the effects of terror on the male birth rate and retreating Antarctic glaciers, is a half-column headed "Ripples cause cosmic doubt." A team from the University of Illinois has been analysing data from WMAP, and guess what? That's right—anisotropy in the CMBR conflicts with Big Bang expectations. Quote:

> Ripples in the faint afterglow of the big bang do not seem to be scattered as randomly as expected. This casts doubt on the theory of inflation, a cornerstone of modern cosmology.

Heaven forbid that observational evidence, of all things, should cast doubt on universal expansion!

Team leaders David Larson and Ben Wandelt scratched their heads, but later admitted that they could not even speculate on what might have caused these deviations. If only they would refer to other studies in the literature, they would find that mapped astrophysical structures correlate remarkably well with anisotropies in the quadrupole and octupole harmonics of CMBR. The raw data from WMAP is open to a wide range of interpretations, and as far as the consensus model is concerned, the analysis is critically based upon the assumption that the Universe is expanding, and with fluctuating rates of acceleration. They explain what they call secondary thermal anisotropies by the Integrated Sachs-Wolf effect, whereby primordial photons from the so-called "surface of last scattering" are energised (blueshifted) by dropping into intervening gravity wells, and then flattened (redshifted) as they climb out of the wells on the near side.

However, because of the supposed expansion of space, there is less blueshift than redshift so there is a net loss in energy (temperature) as the photon proceeds through these massive objects (Note: I am confused by the stated condition of expanding space

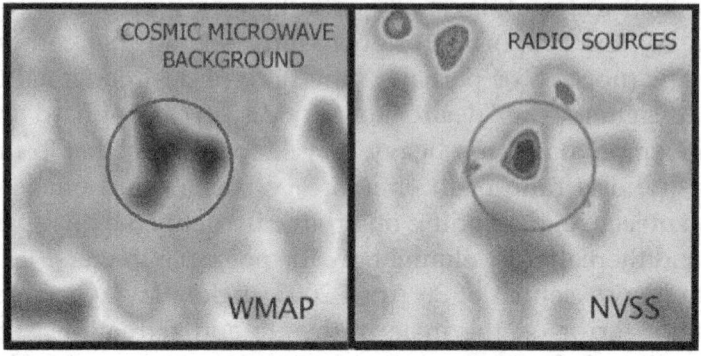

Figure 28: Direct comparison between the WMAP Cold Spot and the radio sky survey image of the Great Void. It is direct observational proof of the link between CMB anisotropies and real astrophysical systems observed in other wavebands. (Comparison image from the website of Lawrence Rudnick, originals courtesy of NASA, Wikimedia Commons, Public Domain).

theory that space inside gravitationally bound systems, that is, gravity wells, does *not* expand, but they conveniently ignore this).

Some photons will travel through more potential wells than others, and so we get thermal hot and cold spots on the microwave sky. Now, we also have to take into account the effect of Dark Energy, which should be the opposite of Dark Matter. In effect, they are claiming that Dark Energy blueshifts light. What do they think this does to the redshift distance ladder? There is alleged to be so much Dark Energy out there that its effect would be catastrophic for any type of systematic Doppler redshift.

A direct correlation has been found between patterns in the WMAP image and the remarkable "Great Void" discovered by astrophysicists in 2007. During the same year, the WMAP team were disconcerted to find an unexpected anomalous pattern in the microwave image that is referred to as "the Cold Spot." Lawrence Rudnick, Shea Brown, and Liliya Williams of the Department of Astronomy, University of Minnesota published a paper in 2007 that correlated the two. It is entitled "Extragalactic Radio Sources and the WMAP Cold Spot" and in it they make the association of astrophysical structure and CMBR patterns very clear.

Rudnick confronts the issue head-on (the emphasis is mine).

> The WMAP cold spot could have three origins: a) at the last scattering surface ($z \approx 1000$), b) cosmologically local ($z < 1$), or c) galactic. Because the spot corresponds to a sig-

nificant deficit of flux (and source number counts) in the NVSS, we have argued here that the spot is cosmologically local and hence, a localized manifestation of the late ISW effect ... our results suggest that the dip in extragalactic brightness and number counts and the WMAP cold spot are physically related, *i.e.*, that the coincidence is neither a statistical anomaly nor a WMAP foreground correction problem. If the cold spot does originate from structures at modest redshifts, as we suggest, then there is no remaining need for non-Gaussian processes at the last scattering surface of the CMB to explain the cold spot... Together with previous work, we rule out instrumental artifacts in WMAP due to foreground subtraction... *the WMAP cold spot therefore would then have a local origin.*

Rudnick was initially sceptical that the cold spot could find a partner in real space that was big enough:

To create the magnitude and angular size of the WMAP cold spot requires a ~140 Mpc radius completely empty void at $z \sim 1$ along this line of sight. This is far outside the current expectations of the concordance cosmology, and adds to the anomalies seen in the CMB.

He soon found the physical alignment, however. Using the NRAO Very Large Array Sky Survey (NVSS), he and his colleagues inspected radio wave images of that region in the sky, and found the very object that could solve the mystery—a vast, 280 Mpc (~1 billion light years) patch of pristine emptiness. They had discovered the Great Void in Eridanus, and it lined up precisely with the WMAP Cold Spot. The interesting conclusion for me is that they now admit that fluctuations in the microwave background are aligned with local astrophysical structure, and therefore not demonstrably a consequence of an expanding Universe. That really puts the cat amongst the pigeons, I'm afraid.[*]

The evidence is piling up against the standard assumption of cosmological origins for the microwave background, it must be said. A very strong indicator that the background radiation couldn't have originated too far away came from a study by plasma physicist Eric Lerner in 1990.[†] Fritz Zwicky inferred from

---

[*] Lawrence Rudnick, Shea Brown, and Liliya Williams, "Extragalactic Radio Sources and the WMAP Cold Spot" *arxiv*: astro-ph/0704.0908.

[†] E.J. Lerner, *Astroph. J.*, **361** (1990), 63. Eric Lerner is the author of a comprehensive, non-technical book expounding plasma theory in a deep social context, called *The Big Bang Never Happened* (New York: Vintage Books, 1991).

observation in 1953 that space between the stars was suffused with gaseous, dusty material. It was later confirmed in detail by South African radio astronomer Gerrit Verschuur. Lerner showed with compelling laboratory-based evidence that the interstellar medium (ISM) would absorb and attenuate radiation at microwave frequencies, and therefore that the CMBR could not have passed through any significant tracts of the ISM. That fact alone effectively invalidates the Smoot interpretation of COBE data. Notwithstanding all those impressively powerful people arguing a kind of "secular divine intervention," we can safely infer from unadorned physics that the IGM radiates at us from all sides at about 3 kelvins.

A combined team from the *Ecole Polytechnique* in Lausanne, Switzerland, and the *Instituto de Fisica de Cantabria* in Spain has been examining the Cosmic Microwave Background (CMB) data from the first year of WMAP, and in 2006, they found what practically every other WMAP study has found: Anisotropies in the CMB are aligned with the ecliptic and/or other local astrophysical structures.

Most analysts stop there, express surprise, and find something else to do. Lead author Y. Wiaux and colleagues don't run away; they take it a step further. They question the supposed isotropy of the rest of the universe based on what we see in the WMAP data. In their conclusion, Wiaux et al. state (my emphasis):

> The present original analysis of structures alignment in the CMB from the first year WMAP data clearly identifies a mean preferred plane... very close to the ecliptic poles axis. [] Various hypotheses can be suggested in terms of cosmological or foreground structures, or systematics. *But nothing at present allows us to discard the possibility of global universe anisotropy, simple violation of the cosmological principle hypothesis.*

Without the Cosmological Principle, Big Bang expansion fails. It's as simple as that. I am steadfastly unrepentant in my assertion that from being originally a much-vaunted "proof" of Big Bang, the CMBR now clearly debunks it.

---

* Y. Wiaux, P. Vielva, E. Martinez-Gonzalez, P. Vandergheynst, "Global universe anisotropy probed by the alignment of structures in the cosmic microwave background" *arxiv*: astro-ph/0603367.

A recent paper by H. N. Sharpe suggests that CMB anisotropy may well come from local structure. I mean *very* local. Sharpe's study finds correlations between CMBR anisotropies and our Solar System. You could hardly get more local than that, in astronomical terms. He says it plainly enough if you can get over the physics-speak:

> A non-cosmological origin for the CMB quadrupole moment is suggested in this paper. Geometric distortions to an otherwise isotropic CMB could be imprinted on the CMB radiation as it propagates through the asymmetric termination shock formed at the boundary of the solar wind and the local interstellar medium. In addition to this boundary distortion, the Voyager spacecraft observed abrupt changes in plasma properties and rapidly fluctuating magnetic and electric fields as they recently crossed the termination shock and entered the heliosheath. Several mechanisms are discussed which could potentially imprint the termination shock distortion on the CMB. Temporal variations of this distortion due to solar wind pressure wind changes could manifest in the multipole moments of the CMB. Speculations are presented for the effect of heliosheath radiative and dynamical processes on the observed small-scale angular power spectrum of the CMB.[*]

I am sorry. I have misled you. Of course there could be a more local source of the microwaves than boundary shocks of the Solar System: The Earth itself! This third rock from the Sun is generally regarded in astrophysics as a blackbody radiating at around 300 kelvins. Pierre-Marie Robitaille (with whom we became well acquainted earlier in this chapter) finds compelling evidence for an Earthly origin of the radiation streaming at us from all sides. As in previously cited instances of alternative explanations to orthodox hypotheses, Robitaille's suggestion is much simpler, uses laboratory physics, has no fudge factors or tuneable variables, and is intuitively more sensible.

Here follows a selection of quotes from his paper that I think summarise a very compelling case for non-cosmological origins of the microwave background. Robitaille freely uses the words of COBE team leader John Mather himself.

> The classic example is given by the Berkeley-Nagoya experiments, just before the launch of COBE. Reflecting on

---

[*] H.N. Sharpe, "A Heliosheath Model for the Origin of the CMB Quadrupole Moment" *arxiv*: astro-ph/0905.2978.

these experiments, Mather writes: 'A greater shock to the COBE science team, especially to me since I was in charge of the FIRAS instrument, was an announcement made in early 1987 by a Japanese-American team headed by Paul Richards, my old mentor and friend at Berkeley, and Toshio Matsumoto of Nagoya University. The Berkeley-Nagoya group had launched from the Japanese island of Kyushu a small sounding rocket carrying a spectrometer some 200 miles high ... The results were quite disquieting, to say the least: that the spectrum of the cosmic microwave background showed an excess intensity as great as 10 percent at certain wavelengths, creating a noticeable bump in the blackbody curve. The cosmological community buzzed with alarm.'

The Berkeley-Nagoya experiment took readings at an altitude of 320 km, and Mather had to deal with a sense of foreboding in his team that higher altitude might not satisfactorily clear the fog. As scientists, they were nagged by doubt. Would the imminent COBE observatory be free of terrestrial influences, or at the very least be contaminated in such a way that allowed *noise* to be stripped away from *signal*? According to Robitaille, the entire microwave signal, or at least the overwhelmingly greater part of it, is from the Earth. The COBE mission was almost fatally compromised by what they considered to be foreground noise, when in fact as objective investigators they should have returned in triumph with a conclusively negative result. Only one COBE result, however, would align with their mission statement, and in pursuing that end, objectivity went out the window I'm afraid.

Robitaille continues, and shows how the COBE officials—in their own words—manipulated the data to eliminate contrary evidence.

> The FIRAS team comments as follows: 'Pending further detailed study of possible instrument faults at these low frequencies, we cannot speculate on their nature. We emphasize that the size of the apparent deviations is greatest at those frequencies where diffractive effects, interferogram baseline curvature, and very low spectral resolving power and wide spectral sidebands cause the greatest difficulties in calibration'. The authors therefore 'conservatively increase the statistical errors by a factor, forcing X to exactly 32, the number of degrees of freedom in the fit.' Nonetheless, they eventually publish new residuals, which have now lost the systematic variations... This shows the power of the fitting methods applied.

Could the radiation, against all expectation, have come from the planet Earth itself? If one weighs up the probabilities, it would appear that it could, far more plausibly than the opposing hypothesis of a primordial fireball. It requires a huge leap in mindset because we are so used to the automatic assumption of an extraterrestrial, extragalactic causal event. It may not have been an event at all, but continuous local radiation. That would explain why the radiation picture changes year-on-year. Theory expects it to be frozen in time:

> The thermal emission of water, in the microwave and far infrared, remains incompletely characterized. Our planet has never been eliminated as the source of the microwave background. In the end, the PLANCK satellite should reveal that the Penzias and Wilson monopole was never present in the depth of the Cosmos. The signal belongs to the Earth.[*]

We will not successfully bring the matter of the CMBR to a useable conclusion unless we first rid the whole affair of an effect I have named Investment Bias.[†] Nobel Laureate Robert Laughlin tells amusingly in his book A Different Universe[‡] of the "First Theorem of Science," attributed to George Chapline: "It is impossible to convince a person of any true thing that will cost him money." Think about it. With what aim and purpose did the sponsors of COBE and WMAP allocate hundreds of millions of dollars and incalculable man-hours?

Clearly, it was to find observational support for Big Bang Theory. There was nothing else on the agenda. Consider this—if by some greatly fortunate quirk of happenstance we had no Big Bang Theory, how would the world of science then have dealt with the microwave fog that surrounds us? Would they have sent up hugely expensive orbiting observatories focussed specifically on obtaining microwave pictures at excruciating resolution?

---

[*] Pierre-Marie Robitaille, *ibid*.

[†] Investment Bias—the subjective prejudice applied to scientific endeavour in order to align results with the outcome mooted in motivation for funding, and thereafter the pressures felt by scientists to maintain that position—is an extraordinarily powerful influence on scientific results, and needs to be clearly defined, recognised, and incorporated both into the literature and into data analysis, along with whichever other biases might skew the results.

[‡] Robert B. Laughlin, *A Different Universe (Reinventing Physics from the Bottom Down)* (Cambridge MA: Basic Books, 2005).

Of course not! They would have treated it as the radio noise it is, and probably left it at that bar a few mildly inquisitive glances. But, my friends, we *can* extract some benefit from the CMBR expeditions, because we have at the very least witnessed with alarming clarity the scientific overkill that went into reinforcing a theoretical prejudice. That ought to be a valuable lesson in itself. Having made the investment however, the sponsors would expect a return. Their business plan demanded it. And so they threw at it everything bar the kitchen sink. No, it is not a sinister, political conspiracy; it is simply economics, and the currency is the dollar, plus, more significantly, exposure of individuals to the diabolical possibility that after so much effort, they might just have been wrong.

The Cosmic Microwave Background Radiation is tremendously important to cosmology in particular and physical science in general. This is not because it is a direct link to creation, but because it is a classic case in our investigation. The CMBR syndrome, if we may call it that, illustrates with eye-watering resolution just why it is we should shy away from the meta-mathematical approach. Like the Hubble law, the political significance of what otherwise should have been seen as benign radiation was hastily extracted from specious evidence, and less contrived fits to mundane physics were overlooked by the far more thrilling prospect of a campaign victory. The radiation background in microwave was canonised and awarded the salutation "Cosmic" before the body was cold.

Then the vast, ravenous engine of abstract mathematical expression was harnessed to the cause by scholars who display the kind of flawed genius we grudgingly admire in computer hackers. They may be chasing the wrong things in life but damn, they're smart! They could, I'm sure, find a replica of the Sistine Chapel in the tealeaves at the bottom of my cup if they were but given the right dose of motivation. Unless we are satisfied that synthetic evidence has a role to play in scientific endeavour, their proven ability to produce results that do not occur naturally is precisely why we should avoid those results like the plague.

I'm going to leave the issue of background radiation there for the time being, but rest assured, the debate is not over yet! Fingerprint of God? No, Dr Smoot, it's just a gentle cosmic landscape, a soft-focus, pastel picture of the familiar, rolling countryside; not a searing flash from the biggest and hottest sub-nuclear blast in history.

> **Discussion:** *Despite chronically flawed instrumentation that could not possibly produce sufficiently accurate data, the COBE satellite was nevertheless credited with measuring "the most perfect blackbody ever recorded in the history of science." The radio fog surrounding us is ambient starlight reflecting local structure and the equilibrium temperature of space. It cannot logically be connected to an expanding Universe or primeval fireball.*

Chapter 7

Structures and Cycles

Social behaviour

Can a static universe resist collapse? What does large scale structure really look like? Observed systems are unlikely given the time frame of the Standard Model. "The only way to avoid this is to go to a cyclic universe model in which the timescale is infinite."

(Geoffrey Burbidge)

✦ ✦ ✦ ✦ ✦ ❖ ✦ ✦ ✦ ✦ ✦ ✦

*G*n the paper "Large Scale Structure in the Universe Indicated by Galaxy Clusters", Neta Bahcall, professor of astronomy at Princeton University, summarises it thus:

> Still, despite the great effort and many ingenious ideas, no single theory for the formation of galaxies and large-scale structure can yet satisfactorily match all observations.[*]

It would appear that observations, at super-galactic scales at least, are always in some or other respect anomalous when tested against the body of theory.

Our descriptive knowledge of galaxies increased exponentially from the time of Hubble's first foray into extra-galactic astronomy in the late 1920s. However, our definitive *understanding* of these systems seems to have simultaneously gone backwards.

---

[*] Neta Bahcall, "Large Scale Structure in the Universe Indicated by Galaxy Clusters" *Ann. Rev. Astron. Astrophys.* **26** (1988), 631-686.

Edwin Hubble designed his Tuning Fork classification system around his belief that galaxies were stable and symmetrical, reducible to a linear hierarchy of just a handful of distinct species. By the 1950s, it was seen that Hubble's galaxy classes were woefully inadequate, and that galaxies were indeed behaving mysteriously. A decade earlier, Erik Holmberg had modelled tidal disturbances apparent in "stellar systems which pass one another at small distances."[*] The subsequent discoveries by Arp, the Burbidges, Struck, and others demonstrated that galaxies were not appreciably diverging, but indeed enthusiastically engaging one another, much as stars do in binary systems. It is at the level of galaxies that the expansion idea meets its nemesis.

In this chapter, we shall deal with structure as it relates to expansion, arranged as the answers to just five questions:

- Firstly, could the structure we see have evolved within the time constraints of the expansion era (13.7 billion years)?
- Secondly, can we detect evidence contradicting universal evolution in look-back time?
- Thirdly, can a non-expanding Universe resist collapse?
- Fourthly, what do supernovae tell us about the rate of expansion?
- Lastly, what is the role of gravitational lensing, if any, in observation of anomalously aligned structures?

The expanding Universe is a peculiar animal, as we have already seen. Structurally, it has some properties that we really ought to take a closer look at, hypothetically speaking of course. First off, it is in the throes of a linear, apparently endless process. From a singular hot beginning in finite time, it expands and cools without conceivable limit. It will endlessly approach zero in density and temperature. We assume that in the fullness of time it will freeze to death in a breathless desert. Meanwhile, it is said to be dynamic in the sense that it is continuously expanding, allegedly at spectacularly varying rates. We, on the other hand, suggest that the Universe is not expanding, and therefore call it static in the terminology of astronomy. It's a bold move, and we are naturally expected to motivate the idea. How could an internally dynamic Universe remain in radial equilibrium overall? Of

---

[*] Erik Holmberg, "On the Clustering Tendencies Among the Nebulae II. (…)" *ApJ*. **94** (1941), 385-394.

Figure 29: Every object—each dot—you can see in this HUDF image is a galaxy. They are mostly high redshift—yet they are not visibly young or nascent. (Image courtesy of NASA, ESA and STScI).

course, we have human limitations; we don't know the answer, but we shall make some suggestions that at the very least use better physics than the Standard Model.

In 2004, the Hubble Space Telescope produced pictures of the most ancient galaxies yet seen. A spectacular array of images, called the Hubble Ultra Deep Field (HUDF), was captured in 800 separate pictures taken on 400 consecutive orbits of the satellite. It is a composite of nearly one million seconds of exposure time, and fully illustrates the fantastic abilities of the HST's imaging instruments.

Notwithstanding the technological accomplishments of the pictures, it is their content that interests us here. Revealed in this stunning panorama, taken from a patch of space below the constellation Orion, are more than 10,000 galaxies from an era that began about 13.3 billion years ago. In view of the fact that the Big

Bang is estimated to have occurred 13.7 billion years ago, these distant galaxies are labelled "very early universe" and are thought to have originated at some time in the first 500 million years of existence. Could mature galaxies have evolved in just 500 million years? It's hardly possible.

If we take the orthodox line and suggest an evolutionary model for the Universe—and remember, I do not—then we need to deal with two evolutionary streams: While the purely gravitational accumulation of material densities and evacuation of voids takes place, a concurrent, entirely different process overlays it. A far more nebulous evolutionary scheme would have to account for the chemical elements, and indeed, also the four fundamental forces of nature that stipulate their existence. For purposes of illustrating the absurdity of the ruling paradigm, I am in this exercise ignoring—with sincere apology—the cosmological roles of electricity and magnetism.

Let us review how galaxies are thought to form, although no one can be certain about this. It soon becomes painfully obvious how little we truly *know* about these hypothetical processes. If we *knew* how old the Universe is, we could test our galaxy formation hypotheses, and if they didn't fit, we would need to rework them. If, on the other hand, we *knew* how galaxies evolved, we could use that to test the age of the Universe. But, unfortunately for science, we don't *know* either of these things; we are left up the creek without a paddle. We can't anchor our cosmology on anything solid so that we can test our ideas. The best we can do is play free parameters off against each other, a crazy tango of ideas with ideas.

It's a multi-tier process in all currently popular models. Galaxies are commonly thought to evolve bottom-up, from star formation in nebulae to clusters of stars, thence to elliptical galaxies, which ultimately spin and collapse into disks around a compact core. If we add up the components, we quickly exceed the posited timeframe. To get to a spiral galaxy out of a cloud of gas and dust is an enormous process, so big in fact that it defies realistic measurement. How long would it take from nebula to disk? How about 10 billion years? That's really conservative—just collapsing the elliptical should take way more than that—but your guess is as good as mine, so let's err on the side of caution. Then there's the process of converting a flattened elliptical to a spiral. For the spiral arms to form properly, the disk needs to rotate at least 40 times. At an average orbital period of about 300 million years (more-or-less the rotational period of our own galaxy, the Milky Way), that's 12 billion years on its own!

Now the really interesting part—how long from the Planck era quark-gluon plasma to a fully fledged nebula? Let's assume that the first nebulae evolved from the particles that formed in primordial quark-gluon plasma. Most analysts say that we are talking hundreds of millions of years, maybe a billion. I suggest it's more likely to be many billions of years because the nebulae would have been collected from a diffuse baryon background, in effect by creating voids. The creation of voids between more concentrated systems would in itself be immensely time consuming. The gravitational action required to suck ultra-low-density material into recognisable clouds interspersed with overwhelmingly vast expanses of emptiness is difficult to get to grips with. In the first place, the attractive force would be vanishingly weak because the nascent cloud is itself only marginally denser than the material it seeks to pull in. Secondly, gravitational potential diminishes as the square of distance from the attracting body; pulling widely dispersed atoms away from their preordained trajectory to form nebulae and leave gargantuan voids is going to be painfully, almost impossibly slow.

If we do some rough calculations using what we think might be realistic streaming velocities, we get answers of around 100 billion years. We need remember that the Lambda-CDM cosmology also predicts significantly lower large-scale velocity flows than are actually observed, and would thus propose even longer time scales for evacuating voids.

At the Second Crisis in Cosmology Conference in Port Angeles, WA, in 2008, an even greater, by many, many orders of magnitude, temporal conflict with Big Bang Theory was revealed. Despite the idea of an expanding universe being around for nearly 100 years, said Lyndon Ashmore of Dubai College in his paper "Hydrogen Cloud Separation as Direct Evidence of the Dynamics and Age of the Universe" there is surprisingly little or no direct evidence to confirm whether the Universe is expanding or not. In an attempt to resolve this issue, Ashmore went through the literature tracing the history of Hydrogen clouds over the last 5 billion years or so. At any epoch in time the average spacing of the Hydrogen clouds should be the same and so in a static universe we would expect to find the same average spacing across the whole range of redshifts. In an expanding universe, the clouds will become further and further apart and so the average spacing will increase as the redshift reduces. The question remains, if quasars are at cosmological distances and Lyα lines do represent Hydrogen cloud separation, then why in an expanding universe are they, locally and on average, equally spaced over a range of redshifts? By extrapolating the data back to a virtual beginning, when the Hydrogen clouds were on average at atomic spacing, we get an age for the Universe of 6×10^{27} years—that is 5×10^{17} times the presently accepted age!

Then there's the thorny issue of galaxy collisions. Studies of supposedly distant galaxies—and don't forget, we are posing a challenge to the distance modulus as it relates to time—show that the largest spiral galaxies are morphologically peculiar, suggesting prior collisions. In a controversial finding, an international team using the Very Large Telescope array in Chile has completely reworked the way that galaxies form. A two-year session using two of the four 8.2 metre telescopes has led to the hypothesis that most of our "present-day" galaxies are the result of merger events rather than the simple gravitational aggregation of stars.

A concurrent but independent survey by the Spitzer Space Telescope has identified numerous strange objects known only as "blobs." They are huge clouds of intensely bright material enveloping distant galaxies, and Spitzer's penetrating infrared vision has revealed that in some cases the blobs hide more than one galaxy. The astronomers believe that blobs may surround multiple galaxies in the process of merging by collision.

That transient galaxy dynamics play a significant role in galaxy morphology is now well established. Collisions or mergers of galaxies are highly prevalent, with nearly half of known galaxies engaged in some stage of physical interaction with another galaxy, and nearly all cohesively-formed galaxies, especially spirals, having experienced at least one, often several collisions in their lifetime. We have already noted (in chapter four) that galaxy collisions militate strongly against expansion theory.

That adds a whole new dynamic to the scheme. In terms of cosmological best-guessing, we've got galaxies that have formed over a period of 120-odd billion years, and they are all flying away from each other at spectacular speeds because of inflation. Now think about it—something has to slam the brakes on, deflect and turn neighbouring galaxies towards each other, let them crash, and finally allow everything to settle down to form a mature, active galaxy. The sheer scale is frightening: Active galaxian nuclei are calculated to have a core mass on the order of 10^6 to 10^9 solar masses—that's an awful lot of material to collect. How can we possibly estimate how long that would take, from start (diverting galaxies towards each other) to finish (settled, mature AGN)?

Shall we now briefly deal with chemical evolution? I shan't make a big deal of this because as we have seen, development of gravitational structure already defeats the expanding Universe. One paragraph from the *meister* of stellar nucleosynthesis, Geoffrey Burbidge, from his 2008 review called "B²FH, the CMB, and Cosmology" will suffice. Let me first explain what B²FH means. It refers to a paper published in 1957 that would become a standard reference in cosmology and related fields. The paper defines stellar nucleosynthesis to this day, and came to represent an entire field of science. It was called *Synthesis of the Elements in Stars* and the authors were Margaret Burbidge, Geoffrey Burbidge, Willie Fowler, and Fred Hoyle—hence, B²FH.

Burbidge addresses the issue of time and the elements as follows:

> B²FH and Cameron in 1957 were able to explain how all of the isotopes in the periodic table with the exception of (four) could have been synthesized in stars. In the cyclic (quasi-steady state) cosmological model, the long time scale means that the other light isotopes, and particularly the high abundance of helium, could have been synthesized as a result of creation in the centres of active galaxies. Thus Oppenheimer's cynical view of the steady state cosmology can be stood on its head. From our standpoint, the observational data, and in particular the energy which must been released in the burning of hydrogen to produce helium, suggests that it has given rise to the observed microwave background. This release has taken place over a long period, too long for a big bang universe. Thus the observational data favour a cyclic universe model.[*]

The matter rests.

I. E. Segal of the Massachusetts Institute of Technology is the author of an alternative scheme he calls Chronometric Cosmol-

---

[*] Geoffrey Burbidge, "B²FH, the CMB, and Cosmology" *arxiv*: astro-ph/0806.1065.

ogy. Although his thesis is undoubtedly interesting, this is neither the time nor the place to divert our attention from the matter in hand, so I shall let Segal make a general summary of large-scale linear evolution and leave it at that:

> Since there is no direct observational means to establish the evolution postulated in big-bang studies of higher-redshift galaxies, and the chronometric predictions involve no adjustable parameters (in contrast to the two in big-bang cosmology), the hypothesized evolution appears from the standpoint of conservative scientific methodology as a possible theoretical artefact.[*]

I doubt I could exaggerate the importance of Segal's conclusion to my general argument that extracting evidence from ideological shadows constitutes nothing less than scientific malpractice.

One last comment on the validity of bottom-up evolution on the cosmic scale: It violates the 2nd Law of Thermodynamics. The second law requires closed systems (like the finite universe) to proceed inexorably towards higher entropy. In simple terms, the 2nd Law expanded to its conclusion states that for every spontaneous interaction, the Universe becomes irreversibly disorganised, something we physicists call entropy. In 1865, author of the 2nd Law Rudolf Clausius put it rather more directly: The entropy of a closed system *increases* with time. The relativistic expansion of the Universe in BBT implies that it is a closed system. In Big Bang Theory, entropy *decreases* with time. I often wonder why there is such a pregnant silence on this issue.

No machine can be totally efficient; it cannot produce more than it consumes. Our understanding of thermodynamics is that inefficiency (entropy) is shown by the production of waste heat. We should, by this law, always move (at the macro scale in a finite universe at least) towards disorder and chaos. We should see the Universe becoming ever less structured and more particulate, and that if this is so, we should find evidence in the form of an infused background thermal radiation. The expanding universe theory postulates the direct opposite of this process—it becomes *more* structured and *less* particulate—and thereby defies the 2nd

---

[*] I. E. Segal, "Is redshift-dependent evolution of galaxies a theoretical artefact? " *PNAS* **96**: 24 (1999), pp. 13615-13619.

Law of Thermodynamics. Let's hear what Arthur Eddington had to say about that:

> If your theory is found to be against the second law of thermodynamics [...] I can give you no hope; there is nothing for it but to collapse in deepest humiliation.

Any realistic model of processes expected from linear, bottom-up evolution of structure produces numbers that exceed the mooted lifetime of the Universe by an absurd amount. Just the expansion itself, being a rapid decrease in overall density, would make the task of developing gravitating centres so much slower that they start to look thoroughly implausible. The whole idea just does not make sense, but then again, was it supposed to? The preceding paragraphs are just an extremely generalised, back-of-an-envelope overview of expansion vs. structure. Let's consider some recent observational evidence before we move on to supernovae.

The first observational test we should apply to the Hubble Space Telescope pictures is to estimate the age of the galaxies on view. If any of them is older than a billion years, then our theory has a problem. The beautifully formed spirals and ellipses dotting the picture make the answer painfully obvious. A thorough technical evaluation of the pictured galaxies was called for, and we got it. The incredible resolving power of the HST had captured images as faint as the 30th magnitude (about 10 billion times too faint for the human eye), and we had an unprecedented glimpse into the darkness of a deep space stellar playground.[*]

At the 208th meeting of the American Astronomical Society, held in Calgary, Canada in June 2006, University of California astrophysicist Adam Stanford announced the discovery by his team of a galaxy cluster estimated to be some 10 billion light years from Earth. Very simply named XMM-XCS 2215–1734, the cluster put yet another nail in the coffin of Big Bang's universe. Stanford admitted that the existence of a mature cluster so long ago prompted a rethink on how galaxies form. In plain language, the structure just cannot be reconciled with a Universe that is only 14 billion years old. Something is wrong—either the observation is

---

[*] The counter-argument that those galaxies are in the near foreground of the picture just doesn't hold up to examination I'm afraid, but that's way too involved to argue here.

misread or the model is incorrect. Either way, we cannot solve the problem by ignoring it.

Whilst Big Bang Theory sits with a mouth full of teeth when it comes to explaining the formation of large-scale structures, there is nevertheless a tacit assumption that galaxies form hierarchically through the merging of smaller bodies into larger and larger systems. Very massive structures are consequently presumed to take much longer to form than small structures, and one can understand the surprise of astronomers when they found well-developed spiral and elliptical galaxies, containing mature stars reddened with age, at the far reaches of the known universe.

One of the rules of thumb used by astronomers concerns the age of stars in various types, or forms, of galaxies. It is well accepted in astrophysics that the stars in elliptical galaxies are more than 10 billion years old, so where does that leave us, now that we have seen elliptical galaxies more than 12 billion light years away? At the very least, our estimate of the age of the universe is seriously skew. It must be older than 13.7 billion years, and if it is, then the other sums in the equation don't add up. The prevalence of reddish-coloured stars in these galaxies is a clear ageing benchmark. A red giant star (which is what our Standard Sun should become before it expires in about 15 billion years) would be typically over 10 billion years old. That means that some components of our "early" galaxies formed at least 8 billion years before the beginning of time.

There are numerous studies indicating developed galaxies at high redshift, with concomitant difficulties for theories who suggest we should see no such thing. There are far too many to list here, so I will mention just two examples. In 2004, H. Chen and R. Marzke announced their results: "Discovery of massive Evolved Galaxies at $z > 3$ in the Hubble Ultra Deep Field." The title says it all.[*] In January 2009, a team of scientists led by Ignacio Ferreras of the University College, London, published the results of their investigation into galaxy structures, entitled "On the Formation of Massive Galaxies: A Simultaneous Study of Number Density, Size, and Intrinsic Colour Evolution in GOODS."[†] One brief quo-

---

[*] H. Chen and R. Marzke, "Discovery of massive Evolved Galaxies at $z > 3$ in the Hubble Ultra Deep Field" *arxiv*: astro-ph/0405432.

[†] The acronym GOODS stands for "Great Observatories Origins Deep Survey". Ignacio Ferreras *et al.*, "On the Formation of Massive Galaxies: A Simultaneous

tation is enough to answer our second question—can we detect evidence of universal evolution in look-back time? They think not:

> We find that the most massive systems... do not show any appreciable change in co-moving number density or size in our data. Furthermore, when including the results from 2dFGRS, we find that the number density of massive early-type galaxies is consistent with no evolution between $z = 1.2$ and 0, *i.e.*, over an epoch spanning more than half of the current age of the Universe.

It could hardly be put more clearly than that.

Central to the idea that the redshift-mapped Universe is simultaneously expanding and evolving from an immature state, is the constraint that functional and morphological evolutionary phases of galaxies should correlate with redshift—the higher the redshift, the more immature the galaxy, as a rule. This is not borne out by observation. Galaxy evolution is correlated with environment, irrespective of redshift.

> There is no shortage of evidence that galaxy properties vary systematically with the environment in which they reside. The distributions of star formation histories, morphologies and masses, all strongly depend on galaxy location, at all redshifts probed so far.[*]

You may be puzzled why I refer so regularly to the utterances and works of scientists in the field with whom I appear to disagree on fundamental issues, so much so that it might even seem that we are foes in the fight for true science. That perception is indeed mistaken; disagreement does not make us enemies. Indeed, it is precisely that polarity of opinion that provides the anvil upon which we hammer out the blade of knowledge. Discordant points of view are the critical precursor to productive dialogue, and without such cut-and-thrust discourse, we remain always on the same spot, marking time, while the sages pronounce Truth. All the rest of us are then merely sycophants whose only visible talent is constrained by our ability to clap hands and cheer at the right psychological moment.

Study of Number Density, Size, and Intrinsic Colour Evolution in GOODS." *arxiv*: astro-ph/0901.4555.

[*] Bianca M. Poggianti, *et al.*, "The ESO Distant Cluster Sample: galaxy evolution and environment out to $z = 1$" *arxiv*: astro-ph/0904.4558.

Such an elder statesman is Alan Sandage, and we disagree strongly on most issues relating to the application of astrophysics to the atmospheric world of cosmology. Nevertheless, I respect Sandage's views because they are so well-considered. He is incredibly well versed in the fundamentals of astrophysics as they are taught to us, and sometimes my jaw drops at his outstanding ability to take so many dispersed data streams into account, all at once, to produce a conclusion that on the face of it is very nearly unassailable. However, it is neither his vast knowledge nor his supreme concentration that lets him down; both are superlative. It is simply the way he was educated, and that can hardly be held against him.

Alan Sandage has produced innumerable written works, including many standard texts for students of astronomy. They do an excellent job—all of them—in passing on to the emerging generation the *modus operandi* he was given in his time of learning. The curriculum is preserved save for some additional, superficial details, and so the show goes on. Unfortunately, as we have seen in the frustration of trying to decipher the properties of our own home galaxy, we cannot properly describe the forest while the trees impede our vision. In 1998, Sandage published a collection of notes to facilitate entrée to the literature for first-year graduates. It is entitled "Observational Tests of World Models." Despite generous plurality in the title, it essentially suggests tests for only one model (which he terms the Hot Big Bang Model) but that is neither here nor there. I recommend it because it gives telling insight to the justifications raised by proponents of the Standard Model, and the formalism employed to engineer supporting evidence.

In the matter of evolution, Sandage takes the predictably standard view that the redshift quantity z is a function of time. I must emphasise yet again that although this is standard practice, it creates a false impression. On the evolutionary axis, many properties of galaxies are studied for any kind of relationship with z, and if it is found, the natural conclusion is automatically that evolution mapped by redshift is given by the results. For example, the Butcher-Oemler Effect is named after two investigators who counted the proportion of blue galaxies to the total for clusters, and found to their satisfaction that the secular morphological evolution of galaxies is given by their complexion—the blue-to-red ratio as a function of age. Of course, this is based on the assumption that the redshift value maps time. But does it?

Let us take an alternative view. If, as Arp and Narlikar seem to suggest, redshift is (in some way yet to be determined) an indicator of internal energy, then the whole picture changes dramatically, and we are entitled to conclude with a totally different, non-evolutionary take on galaxian properties like star formation (indicated by the colour blue). If it is wholly or even partially an energy function, then it fails as calibrator of time, or distance, or velocity. The pendulum of logic swings back towards a cyclical model and away from a finite, linear evolutionary progression that depends upon the assumption of z as a time function.

Evolution as a linear process does not seem likely. On any but extremely narrow, terrestrial biological scales (where it might more properly be termed mutation), and when mapped against redshift on cosmological scales, it fails completely. A far better fit between observation and theory is obtained when theory posits a universal system of local cycles, perhaps in a limitless cascade of size. It is important to note that these cycling processes represent a temporary state of equilibrium, a period of calmness where counterbalancing forces, locally at least, seem to have found their niche in the scheme of things. Dynamically, this peaceful coexistence is expressed as rotation or spin, and static processes occur only within the milieu of greater spinning frames. The Solar System represents a locally settled organisation of rotating objects, and the birdbath on my lawn is an example of statically quiescent arrangements between gravitating objects within a rotating inertial frame.

Here we need to consider something of great importance, which brings us to our third question—can a non-expanding, infinite Universe resist collapse? Does Newtonian Mechanics allow for a Universe that is static? Yes it does. Approach it like this: Localise the rest frame and ask, "Does the Solar System, or the Galaxy, or the Virgo Cluster resist collapse?" It certainly seems that way. Cycles not only allow for a non-evolving, infinite Universe; they also permit it to keep the same approximate spatial distribution of structure. What we see is a progressive organisation of systems in gravitational equilibrium, arranged as a structural hierarchy of ever-increasing size, and having no appreciable or conceivable limit. Mass and kinetic energies balance each other by spinning, and each class of coherent entities forms a component stratum in a larger parent structure, itself rotating and balanced.

In other words, a gravitationally defined, infinite, classically mechanical Universe need not collapse. It is rescued by the hierarchical interdependence of local rotational equilibrium, a cascade of ever larger (or smaller) systems, without discernable limit. I am led to this conclusion by observation. We observe and measure cohesive gravitational systems that are to the limits of our ability to measure in orbital equilibrium. This equilibrium is beautifully described by the Kepler/Newton equations, and demonstrated in the Solar System, galaxies, and clusters.

Given that Newton did not attempt in his science to embrace infinity (and indeed, co-invented differential and integral calculus to try to cope with it), one can by extrapolation say that it does allow a static Universe. If the "static" structures we see and measure are components of ever-bigger (or smaller) structures, each in non-expanding equilibrium *ad infinitum*, then the problem of a mechanically stable, spatially infinite cosmos is solved.*

How could we test such a supposition? The only way, and I mean the *only* way to test a theory is by observation and experience. It cannot be done mathematically. In this case, observation of ever larger and larger structures with no discernable limit empirically supports my suggestion quite definitely.

Conversely, let us examine the Friedmann-Lemaître-Walker-Robertson (FLWR) model. Does observation of the real Universe allow expansion as defined by General Relativity? No it does not. The Hubble expansion as described in GR can only work in an isotropic and homogeneous Universe. Is the Universe in reality such a place? All observation without exception shows anisotropy and heterogeneity in all directions and at all measurable distances in space and time. The theory fails abysmally because of this. It is falsified by observation. And yet it remains the most popular daydream in space science, a globally supported case of chronic wishful thinking.

Now for question number four: What do supernovae tell us about universal expansion? A flippant answer could be "nothing at all," but we should take the matter more seriously than that.

---

* Space constraints allow no more than a footnote here. Please refer to the following paper for a solution to the galaxy rotation anomaly using only Newtonian Mechanics: Charles Gallo and James Feng, "Galactic Rotation Described with Bulge + Disk Gravitational Models" *arxiv*: astro-ph/0804.3203.

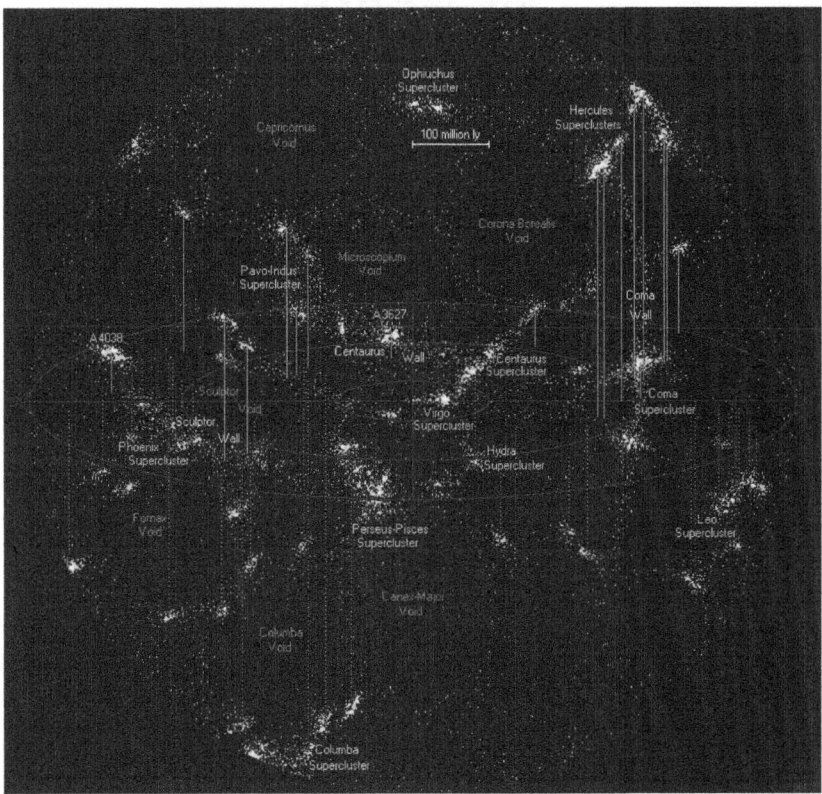

Figure 30: Galaxy clusters visualised in a (supposed) 1 billion light year radius of Earth (in the Virgo Cluster). Neither this system nor subsystems within it show any sign of collapsing. (Wikimedia Commons image by kind permission of the author, Richard Powell).

We need to check theoretical viability, accuracy and representivity of observational data samples, correctness of interpretation, and any occurrence of ambiguity in the results. It's important to emphasise at the outset that I have no interest in fancy theories existing in conceptual isolation; our enquiry should be aimed at deriving an explanation for what happens in the real world, and that immediately requires that we accept the existence of such a thing as substantial reality independent of observation. The stars and galaxies that we are about to discuss are real things, not ideas or models.

To understand cosmic cycles, study explosions. The moment a star dies in a supernova, an inexorable tide of creation goes forth, and it is a beautiful thing to behold. It represents cosmic nativity. A supernova (SN, plural SNe) takes a fraction of a second to explode, yet its brilliance outshines entire galaxies, and the nebula that remains is a starkly fascinating shadow in the picture of galaxies. In that telling instant, redistribution of assets saturates the environment, and consequently, it's so easy to make supernovae major players in theories of cosmic evolution.

There's a problem though. You see, SNe happen far less frequently than the old blue moon—about two observed per galaxy per century. That's not nearly enough—by orders of magnitude— to account for stellar phenomena with anything approaching statistical significance. One per 50 years in a collection of a hundred billion stars isn't going to do much in the bigger picture. But protagonists in the saga of expansion found a use for supernovae that quite exceeds the design parameters for exploding stars. They extracted from observational data a timescale warp in the fading glow of supernovae. Specifically, they targeted those supernovae known as Type 1A.

Convinced that they are standard candles, these devout women and men measured variability in time taken by 1A SNe to fade from their peak brilliance, and concluded with unseemly haste that the differences in apparent duration were not natural properties of varying explosive parameters, but indeed, the effect of expanding space. The idea behind it is that the "light curve"— the graphical plot of brightness varying with time—would be the same for all 1A supernovae if they were measured locally. Measured remotely from Earth, however, the light curves are not the same, and that is unacceptable for standard candles. Explanation: Because they lie at different cosmological distances, the variations in fade duration must be because of expanding spacetime, something known as "time dilation." The immediate conclusion drawn from this interpretation is that all this proves universal expansion. What's more, closer examination, subject to the necessary primary assumptions and fudge factors, indicated to an astonished scientific audience that the rate of expansion was *increasing*. The Universe, ladies and gentlemen, is accelerating away again. So they say…

The real issue here, as I understand it, is whether or not the universe is undergoing systematic expansion, and whether or not SNe rise times (the patterns caused by ebb and flow of luminosity) support that contention. Here's the rub: Do the different light curves not tell us that 1A SNe are in fact *not* standard candles, and that they explode differently over time in each example? That is pretty much how we would normally interpret the observational data in the absence of an overriding theoretical model that tells us otherwise. Unless the progenitor stars of supernovae are geochemically and geophysically identical, we would expect each explosion to plot a unique course on a non-standard timeframe. No one can deny that observable debris fields left after supernovae are so different from one another in so many ways that to suggest the progenitors were all precisely alike is ludicrous. Here again, we are asked by cosmologists to abandon straightforward physics and analyse what we see and measure through *their* spectacles. Do you get an inkling now how annoying that is for us?

The Los Alamos National Laboratory in New Mexico has over the years been home for radically innovative thought. From its days as focus of the Manhattan Project which gave us the first nuclear weapons, out-the-box thinking has characterised the successes of Los Alamos. The legendary Dick Feynman was a citizen there, and so was plasma pioneer Tony Peratt. The most recent news to reach me from Los Alamos is a paper that addresses supernovae light curves in a way that prompts me to say, "Damn! Why didn't I think of that?" The author is John Middleditch, and using SN1987A as an example, proposes that issues still outstanding after 22 years of analysis may be explained by one simple fact, so obvious in hindsight: What we see in supernovae is directly influenced by the progenitor object. He links SNe directly to Gamma Ray Bursters (GRBs) and finds serious issues with the classification of type 1A supernovae as standard candles:

> We note that the bipolarity, enforced on early SN remnants by their embedded pulsars, *i.e.*, very fast axial ejection features within expanding toroids, may complicate their utility, as standard candles, to cosmological interpretation [...] Thus Supernova 1987A, with its beam and jet producing its early light curve and MS, is potentially the Rosetta Stone for three of the four types of GRBs... Since there is no reason to suggest that this is not universally applicable to all SNe, this geometry has grave implications for the use of Type Ia SNe as standard candles in cosmology.[*]

---

[*] John Middleditch "Pulsed Gamma-Ray-Burst Afterglows" *arxiv*: astro-ph/0909.2604.

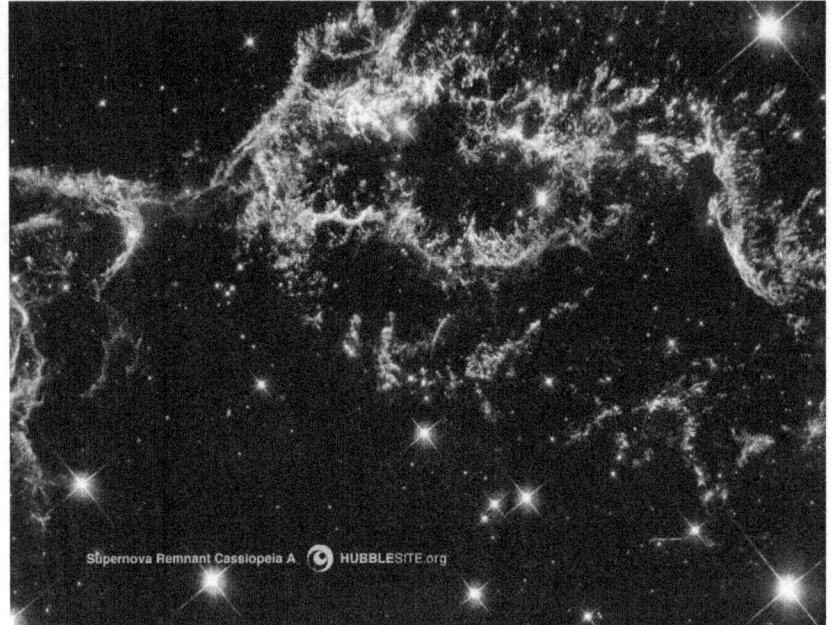

Figure 31: Cassiopeia A, a seemingly chaotic distribution of supernova debris. From a nebula like this, the Earth and the Solar System formed. Are supernovae standard candles? (Image courtesy of NASA, ESA and STScI from Hubble Heritage).

An excellent summary of the principles of SNe light broadening and rise times is contained in Tom Van Flandern's article "Do Supernovas Prove an Expanding Universe?" Quote:

> There is no such thing as a standard single supernova (SN) lightcurve. SN of Type 1a are standard candles only in the sense that their intrinsic maxima are limited to a range of a couple of magnitudes, but this still covers a variety of light curve widths. More importantly, there is no correlation between lightcurve width and redshift beyond that expected from Malmquist bias (the tendency to see only the brightest objects in any class at the greatest distances). The correlation you speak of appears only when brightness is inferred using assumptions about the redshift-distance relation. In fact, it is precisely because the supernovae do not follow the expected behaviour that forces BB proponents to infer that the universe's expansion is now accelerating.[*]

---

[*] Tom Van Flandern, "Do Supernovas Prove an Expanding Universe?" *Meta Research Bulletin* (June 15, 2004).

Jerry Jensen's 2004 paper "Supernovae Light Curves: An Argument for a New Distance Modulus" provides an equally convincing argument that the SNe data contain no evidence of time dilation. [*] This is crucial, because if 1A SNe have no time dilation, that removes the implication that they are moving away from us (or that the volume of space is increasing), and we may therefore with confidence deduce that the universe is not expanding.

Another angle on the alleged relationship between SN light curves and expansion is taken by Australian physicist David Crawford, author of a model called Curvature Cosmology. Early in 2009, he published a paper under the banner "Observations on type 1a supernovae are consistent with a static universe." A strong case can be made, he argues, for a

> ...static universe where the supernovae light-curve-width dependence on redshift is due to selection effects. The analysis is based on the principle that it is the total energy (the fluence) and not the peak magnitude that is the best 'standard candle' for type 1a supernovae. [†]

The death knell for the use of SNe as verification of an expanding Universe was well and truly rung at the 1st Crisis in Cosmology Conference (CCC1) in 2005 by electrical engineer and mathematician Tom Andrews. His paper on supernova light curves was called *Falsification of the Expanding Universe Model and Derivation of the Hubble Redshift and the Metric in a Static Universe*, and as you can imagine, the title alone was enough to get my juices going. The compelling elegance of Tom Andrews' approach is its simplicity. He invokes another class of standard candle, namely Brightest Cluster Galaxies (BCGs), and compares the light signatures with contemporary type 1A supernovae. If the anomalous dimming is caused by Big Bang's postulated expanding space, then the effect should be seen in the light curves of all standard candles, not just 1A SNe. [‡]

---

[*] Jerry W. Jensen, "Supernovae Light Curves: An Argument for a New Distance Modulus" *arxiv*.org: astro-ph/0404207.

[†] David F. Crawford, "Observations on type 1a supernovae are consistent with a static universe" *arxiv*: astro-ph/0901/4172.

[‡] T.B. Andrews, "Falsification of the Expanding Universe Model and Derivation of the Hubble Redshift and the Metric in a Static Universe," in: E.J. Lerner and J. B. Almeida, Editors, *Proceedings of the First Crisis in Cosmology Conference*, American Institute of Physics Conference Proceedings Vol. **822**, 2006).

Although they have the same hypothetical cause, curve broadening and time dilation are different expressions in mathematical analysis. Using two independent sets of data for expansion with a third set (compiled by Goldhaber) as a basis for examining time dilation, Andrews shows that the broadening effect in galaxy light is consistent with neither the expanding universe model nor the notion of time dilation, and in fact directly supports a static (non-expanding) universe model.

In 2009, Andrews put his latest results on the alternative archive *viXra*. The paper is entitled "Discovery of a New Dimming Effect Specific to Supernovae and Gamma-Ray Bursts" and more sensitive readers are cautioned that it contains explicit information extremely disturbing for those clinging to the idea of an expanding cosmos. This 42-page thesis is a comprehensive falsification of the use of SNe data to verify time dilation and expansion. Andrews invokes in addition light curves from Brightest Cluster Galaxies (BCGs) and Gamma Ray Bursts (GRBs) to show that luminosity remains constant during the transient event, thus eliminating both increasing volume of intervening space and stretching of time. Here are some selected quotes (I added emphasis):

> Because type Ia supernovae (SNe) are anomalously dimmed with respect to a flat Friedman Expanding Universe model, it was a surprise to find that the brightest cluster galaxies (BCGs) are not anomalously dimmed. Recently, I found that gamma-ray bursts (GRBs) are also anomalously dimmed... Since the light from the SNe, GRBs and BCGs traverses the same space, the current hypothesis of an accelerated expansion of the universe to explain the anomalous dimming of SNe is disproved. The cause of the anomalous dimming must be specific to the SNe and GRBs... Finally, the light curve broadening effect can be used to determine if the universe is expanding or static. In the expanding universe model, a light curve broadening effect is predicted due to time-dilation for the SNe, GRBs and BCGs. Consequently, if the universe is expanding, two light curve broadening effects should occur for the SNe and GRBs. However, if the universe is static, only one light curve broadening effect will occur for the SNe and GRBs. Fortunately, Goldhaber has measured the widths of SN light curves and conclusively showed that only one light

curve broadening effect occurs. *Consequently, the expanding universe model is logically falsified.*[*]

Tom Andrews has fired a devastating broadside into the bosom of the expanding universe hypothesis. The stark simplicity of his argument simply crucifies the splinter group that came up with the idea that the Universe is undergoing an inflation renaissance. He turned their idea on its head, and used the principles of observational science to thoroughly trounce it. But there is more to it than that, I dare say. Once again in this tale we find that if reinforcement of prejudice is sought passionately enough, evidence *will* be found, even where it did not remotely exist. This is a manifestation of both Investment Bias and Ideological Momentum, the great travesties of modern cosmology.

The link between galaxy collisions and active galaxian nuclei (AGN) is less understood but statistically certain. Collisions are highly disruptive to all components of galaxies, including the nucleus, and the extreme turbulence of AGN suggests the interactive presence of massive gravitational concentrations, possibly multiple neutron stars. It is my belief that concentrations of ultra-gravity are not centred upon Black Holes at all, but on compact objects like neutron stars or systems of neutron stars because that is as far as the compression of matter can go (if indeed it could get that far). Neutron star density is easily high enough to return the mass measured in any known core volume.

A question concerning Black Holes: How could quasars be ejected from galaxian nuclei, given that the cores of galaxies are believed to be Supermassive Black Holes? Big Bang Theory might catch its own tail here, given that the "early universe" would have comprised the most impressive Black Hole ever dreamed up, yet it is blithely credited with ejecting the entire cosmos![†] The question we really ought to ask ourselves is why we have concepts like Black Holes in the first place.

Historically, galaxy counts compiled by Abell (Catalogue of Rich Clusters, 1958), Zwicky *et al.* (Catalogue of Galaxies and Clusters of Galaxies, 1961—1968), and Arp (Atlas of peculiar Gal-

---

[*] Thomas B. Andrews, "Discovery of a New Dimming Effect Specific to Supernovae and Gamma-Ray Bursts" *viXra*:0909.0009

[†] In its early years, the Universe would hypothetically have spent some time within the Schwarzschild radius for total mass present there, and would therefore have presented as a Black Hole. Although this is not acknowledged in the literature, it is implied by the density of the post-singularity "primordial atom."

axies, 1966) made no attempt to reconcile redshift values with other properties in space, but the data were invaluable to later analysts constructing wishfully realistic 3-D interpretations. The Sloan Digital Sky Survey (SDSS) and the Centre for Astrophysics (CfA) survey, as two examples of modern works, have given us 3 dimensional interpretations of pie slices of the universe that rest, or fall, with redshift distance.

All the mentioned surveys produced peculiar patterns when arranged spatially according to redshift, and even more obvious anomalies where resolution permitted detection of material connections between bright objects. Several studies have tried to apply the redshift value of objects in sky surveys in order to establish remoteness, in other words, their depth along the z-axis.[*] The results were weird, at least according to LCDMM, which expects no expression of large-scale structure or preferential location. The spatial distribution of astrophysical objects aligned by redshift is so extraordinary that if true, requires a cataclysmic revision of cosmological theory.

The first shock came from a study that resulted in the notorious "Fingers of God." J C Jackson in 1972 found an observational effect in galaxy distribution data that caused clusters of galaxies to appear elongated when expressed in redshift space, taking on the appearance of "fingers" pointing towards Earth.[†] The virial association of high velocities in clusters with their gravitation distorts the Hubble redshift relationship, and consequently, distance measurements are inaccurate, that is, anomalous according to the model. N Kaiser in 1987 revealed a related but smaller effect occurring in even larger structures. These "Pancakes of God" are attributed to line-of-sight distortion unrelated to distributions predicted by the virial theorem. They are thought to arise instead from infall motions of galaxies as the cluster forms, based on the assumption that high-redshift objects are nascent.[‡] What are these structural simulations telling us? Halton Arp put it neatly: "The

---

[*] See this paper for a description of what may be the largest redshift-defined structure yet discovered: John. G. Hartnett, Koichi Hirano, "Galaxy redshift abundance periodicity from Fourier analysis of number counts N(z) using SDSS and 2dF galaxy surveys" arxiv: astro-ph/0711.4885.

[†] J.C. Jackson, "A critique of Rees's theory of primordial gravitational radiation" MNRAS 156 (1972), pp. 1-6.

[‡] N. Kaiser, "Clustering in real space and in redshift space" MNRAS 227 (1987), pp. 1-27.

fingers pointing at us are telling us our assumptions about red-shift are foolish."

Now, I am not at all suggesting that we do not occupy some privileged place in the Universe; it is so, if only because the Earth is where we uniquely prosecute our point of view. What I am saying is that standard redshift theory applied to the alignment of large scale structure clearly violates the declared principle of Lambda-Cold Dark Matter cosmology that insists that there is nothing special, anywhere. My function in this summary is to point out where standard (consensus) interpretations produce curious results, and in this case specifically, as a consequence of introducing redshift-distance to our calculations of large scale structure.

Whatever the cause, we have a peculiar picture being painted, and one wonders whether that is what actually happens out there, or whether we are just playing games with symbols, terms, and operators. These structures contradict the Cosmological Principal yet again, indicating that the Earth indeed occupies a special place in the Universe. In fact, some analysts take it one logical step further: Redshift-aligned structure at the large scale suggests tantalisingly that we are geographically at the *centre* of the Universe! After that, shouldn't we hand it over to theologians?

Either the redshift data are anomalous, or the implied spatial properties do not fit, or both. They are anomalous also for the λ-CDM model. In the paper "Large Scale Structure in the Universe Indicated by Galaxy Clusters,"[*] Neta Bahcall states:

> The large scale structure results discussed in this review, however, constitute a difficulty for CDM. Considerable evidence for structure on scales $\geq 30h^{-1}$ Mpc has now been accumulated by a number of investigators; this large-scale structure (and velocity) cannot be matched by unadorned CDM models...If these largest scale results are confirmed by new and deeper observations, it will be damaging to the simplest CDM models.

In chapter one, I pondered the boundaries that should limit sensible discussion on the issue of universal expansion, and came upon a quandary. Everything in the Universe seems to be linked in an endless Machian web, so really, any astronomical observation at all is in some way relevant to its participation in the cos-

[*] Neta Bahcall 1988 "Large Scale Structure in the Universe Indicated by Galaxy Clusters" *Ann. Rev. Astron. Astrophys.* **26** (1988), pp. 631-686.

mos, expanding or not, and has something to say about it.* It might be helpful to us to examine one last, particularly blatant example of cascading *ad hoc* physics that empowers the paradigm. So, if you will allow me just the slightest touch of latitude as we wind this arduous journey down to its conclusion, I shall briefly address the issue of *gravitational lensing*.

The idea was first properly articulated by Albert Einstein in discussing the cosmological implications of General Relativity. It's simple, really—the path of a photon is bent from true by the proximity of a massive object. A foreground object, a galaxy for example, could conceivably act as a lens, bending the passing rays from another distant background source around itself in such a way that an image is brought to focus in *front* of the intervening galaxy. If we Earthlings were perhaps located closer to the lensing galaxy than the focal point, we might see a "ring" or multiple identical images of the lensed object. This phenomenon is in fact also predicted by Newtonian Mechanics, although with Newtonian "half-deflection," the lensed object would obviously have to be much further away. In theory, this is all quite straightforward; in practice, it is quite another matter.

Gravitational lensing is one of the tools of the expanding universe set thrown willy-nilly into the fray to counter arguments for a static Universe, and commonly raised against Arp's quasar associations. The quasars "encircling" Seyfert galaxies are proposed as multiply-lensed images of a single remote background quasar. Arp and López-Corredoira have independently found that the suggestion is fallacious, in the cases of their own observations at least.

> Weak gravitational lensing by dark matter has also been proposed as the cause of the statistical correlations between low and high redshift objects, but this seems to be insufficient to explain them, and cannot work at all for the correlations with the brightest and nearest galaxies... (they) have not solved the question of the correlation of galaxies and QSOs, because cross-correlations with bright

---

* I use Ernst Mach's name in a general sense to illustrate the idea of universal coherence. Mach's Law is more specific, attributing inertia to the pervasive, endless effect of gravitation.

and nearby galaxies, which are the most significant, are still without explanation in standard cosmological terms.[*]

Nevertheless, the idea is pursued vigorously in the literature, and a study from the University of Sydney is an example. Several questions in respect of microlensed quasars raise themselves to the objective investigator, especially that the multiple images should all have identical optical properties if they are indeed renditions of the same source object.

The authors seem to suggest an answer to why they are not:

> However, while the combination of deflections due to individual stars [in the lens] is linear, the resulting magnifications are highly non-linear, leading to significant computational challenges…

They state further problems:

> Several quasar systems seem to possess anomalous flux ratios… meaning that observed image brightnesses differ significantly from predictions drawn from gravitational lens models…

Predictably, they solve their quandaries with Dark Matter:

> Two key hypotheses have been put forward to explain these observations; either these are anomalous ratios due to millilensing … by clumps of dark matter in the halo of the lensing galaxy, or the quasars are microlensed by stars embedded in an overall smooth dark matter distribution.

Please note the introduction here of "millilensing." We shall return to this shortly.[†]

A note of caution: You will recall from chapter five that Luyten discovered the high apparent lateral motion of quasars in the late 60s, and tables of data amplifying the evidence were compiled by Hewitt and Burbidge in 1993, and later still by Y. P. Varshni. This is of paramount significance to the assessment of lensing candidates. If you try to visualise the geometry of gravitational lensing, you will quickly realise that the background and foreground objects have to line up with precision, such taxing precision in fact that Fred Hoyle calculated the chances of their

---

[*] M. López-Corredoira and C.M. Gutiérrez, "Research on candidates to non-cosmological redshifts" in: E.J. Lerner and J.B. Almeida, Eds., *1st Crisis in Cosmology Conference, CCC-1*, (AIP Conference Proceedings Vol **822**, 2006).

[†] Hugh Garsden and Geraint F Lewis, "Gravitational Microlensing: A parallel, large-data implementation" *arxiv*: astro-ph/0907.0068.

Gravitational Lens G2237+0305

Figure 32: The Einstein Cross, assumed in its name to be gravitationally lensed renditions of a background quasar, with suitable
further modification by gravitational microlensing. A simpler and
more obvious explanation is that it is a family of quasars around
an active galaxy. (Wikimedia Commons public domain image
courtesy of NASA, ESA/Hubble, and STScI).

fortuitous alignment at 1 in 500,000. Each of the nearly 30,000 instances of excess quasars around galaxies studied by Halton Arp
cannot be attributed to lensing for very long, if at all.

High proper motion of quasars means that the delicate, precise longitudinal alignment of quasar and galaxy cannot maintain
its integrity for more than a decade or two at most. The "Einstein
ring" will magically disappear as soon as line-of-sight integrity is
broken by the transverse motion of the quasar; the duplicated
images will be here one minute, gone the next. It shouldn't take
long. Fortunately for us, we have precise measurements of quasar
systems going back more than 40 years, and lensing simply hasn't
survived the test of time. Every single one of Arp's observed systems is still out there for us to see.

The heath around this particular bush is well and truly
beaten, but Halton Arp tackles the matter directly. He uses a classic example commonly called the Einstein Cross, a system of 5 related objects labouring under the proper name G2237+0305. If
anyone interested in cosmology has not yet read Seeing Red—
Redshifts, Cosmology, and Academic Science, they should do so

without delay.[*] With not a trace of hysteria, Halton Arp cogently reveals the fallacies and iniquities of modern scientific practice, and he does a particularly good job on the Einstein Cross. Please refer to pages 172 to 175, commencing with the section headed "Distribution of Quasars around Seyferts." John Huchra, who discovered the Einstein Cross in 1984, admitted in conversation that his first thought was the chilling likelihood that Arp had been right all these years.

The Einstein Cross has become the best publicised example of gravitational lensing to date. I have to admit it is spectacular, if for no other reason than that the formation is so clearly symmetrical—four satellite quasars of identical redshift surrounding a central galaxy like Beefeaters guarding the Tower of London. In Arp's words,

> Gravitational galaxy lensing just had to be invoked for this one. Subsequent high resolution observations showed four quasars of $z = 1.70$ in the form of a cross approximately centred on a 14th magnitude galaxy of $z = 0.04$. Since the four quasars were all within one arc sec of the galaxy nucleus it was impossible to claim accident. But the gravitational lens was in big trouble from the start because Fred Hoyle quickly computed that the probability of such a lensing event was less than two chances in a million!

Halton Arp, like me, is somewhat sceptical of gravitational lensing, and examined the released NASA image very closely. "I looked at the picture," he says, "and saw what should not be there, a luminous connection between one of the quasars and the nucleus!" It was a smoking gun. He immediately took the picture to his trusted friend Phil Crane, a member of NASA's instrument definition team for space telescope imaging who had access to all the right tools and raw data. A little work overlaying isophotes on the picture soon revealed an even more sensational fact: Not only one, but *all four* quasars were physically joined to the parent galaxy by matter bridges.

The notion that the Einstein Cross was a gravitational lens was crushed; it was simply physically impossible. Even amongst the most enthusiastic promoters of gravitational lensing there were those that were quietly unconvinced, and one in particular took it a step further: He uncovered the hard physical evidence

---

[*] Halton Arp, *Seeing Red: Red shifts, Cosmology, and Academic Science* (Montreal: Apeiron, 1998).

for bridging. Safe from political fallout and prying ears in the shelter of a quiet corridor in Buenos Aires, leading astrophysicist and Big Bang aficionado Howard Yee felt compelled to confess.

Walking next to Halton Arp en route to an IAU conference session, Yee confided,

> We put the slit of the spectrograph between quasars A and B in the Einstein Cross … between them we found a narrow Lyman alpha line—it looks like there is low density gas … between them.

Seen in the light of alignment of A and B with D and the nucleus itself, it was electrifying proof of a physical connectedness over the system as a whole. Crane and Arp's isophotes had tracked the plasma filaments unifying the satellites and the core.[*]

Below a contoured picture of the Einstein Cross on page 174 of Seeing Red, Arp makes a chilling observation.

> One should carefully consider the following important question: What is the chance that a person who notices a discrepancy in a scientific announcement has the opportunity to check it out at the level of the primary data?

The chance is about zero, and we should be giving thanks on high for the extreme happenstance that allowed Chip Arp to dig the truth out of suffocating propaganda. But the scandal was not yet over. The proprietors of universal expansion were not about to roll over and die.

It was imperative that this evidence be published loudly and clearly so that science could redirect itself along a sounder path. Halton Arp and Phil Crane wrote up the results in the appropriate format, and submitted their report to Astrophysical Journal Letters. The editor forwarded it to a referee who had just published a paper supporting lensing in the Einstein Cross. He rejected it. Arp appealed. It was sent to a second referee with even greater bias towards the orthodox model. He implied that since it disagreed with current theory, *the observation must be wrong*, and promptly rejected it as well.

Arp made his dissatisfaction plain.

---

[*] Isophotes in astrophysics are analogous with isobars in meteorology, or contour lines on relief maps. They connect points of equal magnitude (brightness) on maps of astrophysical systems, and are invaluable in tracing physical connections between objects.

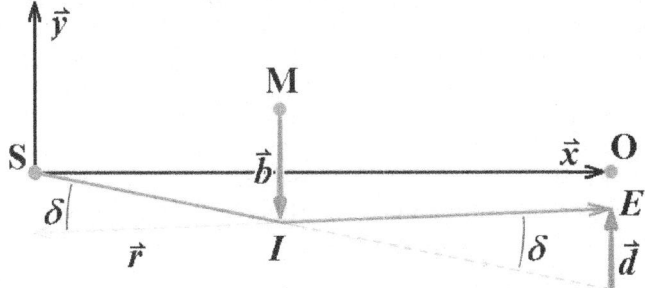

Figure 33: A diagram of the alignment of components in a gravitational lens by Louis Marmet. Applied specifically to the Einstein Cross, it shows that only two images are theoretically possible, not four. (Image courtesy of Louis Marmet from his web page)

> I feel very strongly about what happened, and I want to make my position clear: Astrophysical Journal Letters is the normal journal for publishing new observations from the Hubble Space Telescope... The overriding, first directive of the editor is to communicate important new astronomical results. If the editorial process violates its primary responsibility, it misuses public funds.

It seems the editor in question valued other things in life. Astrophysical Journal Letters never did publish the report.

At the 2nd Crisis in Cosmology Conference (CCC2) in the picturesque harbour town of Port Angeles, Washington State, I was delighted to make the acquaintance of Louis, son of my former mentor, eminent Canadian classical physicist Paul Marmet. Louis has with great aplomb taken the baton of reality physics from his late father and is running with it. On his website, he poses some searching questions around the Einstein Cross, and the verity of lensing in this case is again called into dispute.

The problem Marmet highlights is again one of alignment. We see what is supposed to be four distinct renditions of a single background quasar around the foreground galaxy core. The question he asks is, how could this happen?

In his own words,

> The Einstein cross is intriguing because a mass having a spherical symmetry can only give rise to images located on a straight line. If the mass is concentrated in a small volume, only two images are produced by the gravitational

lens ... A ray of light emitted by a source S is indicated by the blue line. The ray is bent at point I by the mass M and reaches point E ... For the observer O to see the source, the point E must coincide with O ... Therefore, the observer can only see an image of the source S located somewhere on the y axis. Careful examination of the equation shows that there are only two solutions, producing an image of S on either side of the mass M. This may explain two of the four images of the quasar in the Einstein cross.

The other two, if indeed they are the product of gravitational lensing, would require such fortuitous and conflagrated alignment as to render the whole idea improbable in the extreme. But really, why are we arguing about this? Surely Halton Arp makes better sense by proposing that the quasars are part of a physically cohesive galaxian nest. By the principle of Occam's razor, we should ignore the idea that lensing explains the observation, in this case at least.[*]

The relentless cascade of emergency fixes permeating cosmological theory is exemplified without peer in the case of gravitational lensing. To the already questionable application of gravitational lensing to even the most unlikely of candidate systems, the subtle art of *microlensing* (additional sub-lensing by individual stars on the near side of the lensing galaxy) was quickly introduced.

They had no choice. Something was needed to explain away the anomaly of remarkably proximate alignment of quasars around central galaxies, and at the same time, allow for the often different optical properties of supposed "lensed" renditions of a single background object. When microlensing was found to be inadequate, the idea was stretched to *millilensing*. Whither next, I wonder? I'd advise these mental contortionists to simply call the whole show "Dark Lensing" and be done with it.

Gérard de Vaucouleurs was a realist upon whose scientific attitude I have unashamedly modelled my own approach. Throughout his enormously productive life, he did not waver from the path of empiricism, and set an example that has been sadly ignored to this day by the orthodoxy. We recall with wistful

---

[*] For an additional, more technical analysis of the lensing possibilities of the Einstein Cross, see Louis Marmet's web page
http://www.marmet.org/cosmology/einsteincross/index.html.

fondness his brave attempt with Alan Sandage in 1958 to classify the observed manifold of galaxian morphologies.

In a 1970 article in the popular journal Science,[*] an obviously world-weary de Vaucouleurs warned us sombrely of

> ...parallelisms between modern cosmology and medieval scholasticism... Above all I am concerned by an apparent loss of contact with empirical evidence and observational facts, and, worse, by a deliberate refusal on the part of some theorists to accept such results when they appear to be in conflict with some of the present oversimplified and therefore intellectually appealing theories of the universe.

Of course, the ghastly possibility that the model is falsified by observation is never brought into consideration. Thomas Kuhn was right; come what may, the paradigm is sacrosanct.

Discussion: The structural properties of the cosmos militate against the finite, expanding model, and favour an eternal, static, cyclic Universe. The array of technical fixes introduced to counter the evidence of plain sight is unconvincing and contrived, no more than desperate glue in the joints of a globally worshipped house of cards.

---

[*] G. de Vaucouleurs, *Science,* **167** (1970), p. 1203.

Chapter 8

An Expanding Mind

I think, therefore it is

An historical, ontological, and epistemological review of how geometrical curvature came to dictate observation. "No amount of observations will be able to decide on the true geometry of the Universe."

(Oxford cosmologist Joseph Silk)

✦ ✦ ✦ ✦ ✦ ✦ ❖ ✦ ✦ ✦ ✦ ✦ ✦

The late Tom Van Flandern once said, "If you want to find evidence refuting Big Bang Theory, just point a telescope at the sky." Unfortunately, no one did. At least, nobody involved in writing up modern cosmology did. It came from a set of differential equations that very few people in the whole world could claim to understand. There wasn't a telescope involved at all.

It's time for us to have a solemn chat. In this penultimate chapter, I am going to make a very necessary digression from the strictly empirical account this book claims to be (and for the most part, is). I know it's asking a lot, but please bear with me. It is crucial that we clearly understand the justification raised for the expansion of space, and that requires your patience and concentration for the next several pages. There is little chance that we can make progress without a clear historical and epistemological review of how geometry came to dictate observation. I shall try very hard to make the journey worthwhile. Of course, if you are

amongst the fortunate few already done with the idea that imagined curvature constrains how we should measure the cosmos,
then there is no pressing need to read this chapter.

So that we are on the same page here, let's start off by defining the word "space" as we shall be using it. The mathematical
definition of space is "a set of elements or points satisfying speci-
fied geometric postulates." No, we shan't be using that one.
Space, in simple terms preferred by reality physicists, is the seemingly boundless volume surrounding us in which real objects
manifest themselves. It is intrinsically empty, inert, 3-dimensional
(height, breadth, and depth), and featureless. In this work, we as-
cribe no virtue to space other than its potential to contain phenomena. It's simply the limitless place that would be an empty
chasm were it not for the creatures of the cosmos. Aether or no
aether, it's the same everywhere, I believe, and treats us all
equally. As a parameter of measurement in space science, therefore, it may be left out of our equations. Now that's where we hit
serious problems with consensus cosmology.

The difficulty we seem to experience is in differentiating between formulae symbolising the form, and the form itself. We
shall be talking here about the *geometry of space*, and to even contemplate such a thing, the original postulates (and indeed, axioms) of this field of endeavour had to be so substantially transformed as to render them useless in the bigger scheme. The revolution was, of course, driven by physical need; mapping a curvi-
linear surface (such as that of the Earth) could not economically
be accomplished by plane geometry. The pioneers at that time
produced a solution that worked, no question about it, but it incorporated an abstraction that allowed extrapolations to run riot.

There are three guidelines I try to follow in the matter of
space curvature. They are: As a matter of principle, we should use
the most simple conception of space that allows us to make sense
of what we observe; as a matter of fact, there are no *observations*
(that I am aware of) that cannot far more easily be understood in
a Euclidean-space-plus-time-axis framework than in a curved-
spacetime framework; finally, the attempt to force "spacetime
curvature" and "expansion" onto the universe is quite contrary to
what I like to call sensible science, and more importantly, to what
we observe.

It's going to be bitter pill to swallow, but we have to admit
that a piece of mathematics written down to help resolve a prob-

lem we have in the world about us does not actually describe any part of that world. It merely establishes relationships and uses units of measure to reveal useful patterns in the quantities of things. The number of kilometres that you and I are at this moment apart tells us nothing at all about nature per se, but it is enormously helpful in estimating, for example, how long the travel time of a visit would be. Mathematics can be invaluable useful in helping us apply our physics to the world, but it is not part of that world, nor indeed should it in my view be considered part of physics. So, when mathematical theorists talk of "the geometry of space" they are implicitly talking as if space is something tangible with a shape that can be discussed. Of course, mathematicians, particularly those who tackle the enormously challenging problems inherent in cosmology, really do know what they're talking about when it comes to geometry itself; it's just that most of those ideas having nothing, or very little, to do with physics. And this goes to the heart of the problem that this book is attempting to address.

The notion of universal expansion arrived at the doorsteps of startled physicists nearly one hundred years ago. In order to fully grasp the nature of the revolution that was about to take flame and raze the bastion of Newtonian Mechanics, and with it, the importance of the sense of sight to celestial theory, we need to step carefully back even further into the history of scientific thought.

For millennia, humankind had manipulated the properties of points in space by means of geometry, literally, "measuring the Earth." That's where it started: The Earth. This vast planet, so impossibly huge on the paltry scale of people, is where the science of lines, angles, and ratios representing objects in our field of view was born. It came out of agricultural necessity on the river banks of Egypt.

The initial geometrical measurements by land surveyors and architects assumed a flat Earth as the rigid plane underpinning constructions. Locally, curvature was insignificant, and we need look only as far as the pyramids for confirmation. One can build a perfectly good pyramid on a sphere as long as variation from flat in the foundation is trivial, but we should not be lulled into false bravado. Notwithstanding any great and ancient monument, this question of curvature would come back to haunt us. It would drastically alter the course of history.

The rich fields of geometry would, in time, extend far beyond the strain of surveyors' ropes. It was the great Greek philosophers of that marvellous era from the 6th to the 2nd centuries BC who raised geometry to the abstraction of pure mathematics, and allowed scholars to map things that were so far away, they were actually out of sight.

Here is the principle: By looking carefully at things that were nearby, these visionaries could infer those that lay beyond the horizon. They deduced that the Earth was spherical, and set about measuring it. We should note an important corollary to this principle before we proceed. The nature of space itself was not at that time taken into account in geometrical measurement. Space was simply accepted as a universal, unchanging parameter consistent in all directions, and affecting all material phenomena equally. As we shall soon see, this was some 2,000 years later to change dramatically, with far-reaching consequences for astronomers. For the pioneers first measuring the size and shape of the Earth, however, calculation was unsullied by such ideas.

After some effort, it was done: Erasthenes in 240 BC used a comparison of the shadows of two sticks at noon on the Summer Solstice, one in Alexandria, and the other some distance directly south at Syene. It was brilliant and meticulous; he arrived at a radius for the Earth of 6,400 km. That's as near as dammit is to swearing to the figure we have today. It was a truly remarkable achievement.

The man who defined the art was a professor at the Great Library of Alexandria named Euclid (c. 330–275 BC), and he did so in spectacular fashion. My admiration for him is boundless, so it will not do to get me going on the subject. That's another book, another time and place. Suffice it to say that his classic, The Thirteen Books of the Elements, lays out fundamental tenets of geometry so that it can be applied to our physical environment without manipulating dimensions.[*]

Everything was fine, it all worked perfectly well, and there was no pressing need to change it. Ah, what fertile ground that is for curious intellect! Never mind necessity, just the fact that it could be challenged was reason enough. Let's be certain of one thing: The nature of space has not changed from that time to this.

---

[*] Euclid The Thirteen Books of the Elements. My copy is the 1908 translation by Sir Thomas Heath (New York: Dover, 1956).

Nor, from what we can see, does it change from nearby to far. It is the same. The space in my room and the space between the stars can be usefully addressed by the same old set of rules. What has changed, and significantly so, is how we think about it officially.

Carl Friedrich Gauss (1777–1855) is remembered as the Prince of Mathematicians, and it is he who holds a close second place in my adulation, after Euclid. Gauss lived a simple life. Whilst engaged in conjectures of stupefying complexity, he did not, like so many mathematicians of his time, lose his mind or his health on the altar of superhuman concentration.

Astronomy and mathematics have always been comfortable bedfellows, and for centuries leading mathematicians were enticed into putting their formulae to the heavens above. Gauss was no exception. He was for many years the director of the observatory at Göttingen, and legend has it that during all that time, he only thrice spent a night away from it. His students revered him for his even temper and generous spirit, and all in all, we couldn't have wished for a better professor, except for one critical weakness: His obsession with the abstract. Carl F. Gauss just couldn't resist the temptation to indulge and favour his imagination when it was challenged by reality.

For over two decades, Gauss was under commission to the House of Hanover to produce a comprehensive geodetic survey, and today we speculate whether that might have been the inspiration for his revolutionary ideas on topology. Soon enough, the problems of mapping on the surface of a sphere were staring him in the face. It was clear from the outset that mapping curvature with Cartesian co-ordinates in a Euclidean framework would be so clumsy that radically new techniques were called for. And that, my friends, was right up his street.

Gauss generalised the strictures of curved surfaces and presented his scientific progeny the gift of Differential Geometry, eventually used by Albert Einstein to famously define gravitation in the 1915 Theory of General Relativity. As far as I know, we may thereby fairly blame Carl Gauss for stimulating the advent of a consequent field of endeavour which became known by the oxymoron "mathematical physics."

From here on, right through to the advent and application of General Relativity in the scientific revolution of the 20th century, the core issue was curvature. Can space, even in theory, have a shape? For Professor Gauss, acting and thinking as a mathemati-

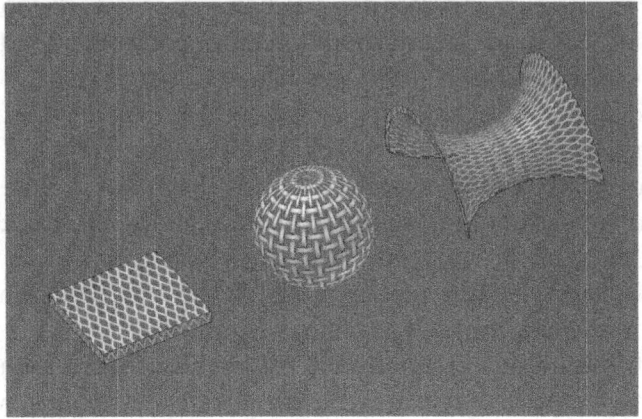

Figure 34: Euclidean, Riemannian, and Gaussian curvature. (Diagram by Gregg Barlow)

cian engaged in the application of terrestrial geometry to the cosmos, the answer seemed to be "yes!" The orb of the Earth, it was reasoned, is mirrored by the celestial sphere, and a horizontal line, strictly speaking, is curved.[*]

Recently, I had technical assistance with concepts in this book from a very highly qualified mathematical theorist, a man professionally engaged in research into General Relativity. Our editorial dialogue, and probably our friendship, came to a sticky end on this very issue: Is a horizontal line curved or is it straight? To me, it's perfectly obvious—a horizontal line is concentric with the surface of the Earth, and is therefore an arc, a segment of a circle. It's curved. My friend responded that it appeared curved to me only because I projected it mentally into Euclidean space. If I saw it in Riemann space, he said, it would be as straight as a die. But, I sputtered impatiently, I'm not projecting it anywhere! The surface of the Earth is the surface of a sphere, and that's curved, period. I lost my cool. That's unforgivable. I'm sorry, but I do find it hard at times.

Gauss reacted to the problem with typical forthrightness. He developed a theory of surfaces to accommodate curvature, essentially to "straighten it out," as it were. Gauss's surface topology

---

[*] If a horizontal line is constructed with a spirit level or modern equivalent as builders do—placing pegs so that the tops are dead level with each other—the line will reveal natural planetary curvature over large scales or with high-resolution instruments on local scales like building foundations.

was not initially a theory of spaces as such, and it's hard to pin-point exactly where surfaces in space became surfaces of space. Nevertheless, his work was enthusiastically supported by one of his star pupils, the intense and frail Bernhard Riemann. What emerged from these meditations was the division of geometry into three subsets: The original Euclidean doctrine, which as a re-sult of these divisions became known as parabolic geometry; then the concave plane of Gauss (hyperbolic); and finally the convex plane of Riemann (elliptic).

These frameworks appeared to have changed the fundamen-tal nature of straight lines, and allegedly altered, as examples, the properties of triangles and parallel lines. The definitions, axioms and postulates depended upon by Euclid were undermined, and chief offender, it was said, was the parallel postulate. In elliptic geometry, the included angles of a triangle summed to more than 180°, and even more astonishingly, parallel lines could meet. A pair of mathematicians had in short order fundamentally changed our conception of reality.

Examples of this odd new world were cited *ad nauseum*: Sup-posedly parallel lines of longitude meet the poles, and the sum of internal angles of a large enough triangle traced upon the Earth's surface did indeed exceed two right angles. It was all because of curvature. Mathematical historian Herbert Westren Turnbull[*] drew an analogy with a strip of sticking plaster[†] on a human hand. Stick it on the back of the hand, and a triangle drawn on the plaster is perfectly normal, in the Euclidean sense. Place it over a knuckle, however, and everything changes. The plaster stretches, resulting in a triangle that exceeds 180° (Riemannian el-liptic). Press the plaster into the depression between knuckles, and the triangle distorts by being compressed into a saddle, and the angles fall short of 180° (Gaussian hyperbolic).

The critical error, I believe, in the thinking of Gauss and Riemann—and all those that followed their amazing meta-geometry—was that conceptually, they let *space* be represented entirely by a particular *surface*. In a nutshell, what they had done is decree that geometry be raised above the measurement of physical objects occupying space, to the level of formulae defin-

---

[*] In James R Newman, editor, *The World of Mathematics*, Volume 1 (Toronto: General Publishing Company, 1956).
[†] Adhesive bandage.

ing the very nature of space itself, a field so incomprehensibly great—and so intangibly abstract—that it encompasses, articulates, and indeed, *exceeds* the entire Universe.

Regrettably, these ideas have become so embedded in modern theorising about the cosmos that we are expected to accept that curvature is inherently embedded in physical reality for no other reason than that it has been so defined. The attitude expected of scientists in this regard is rather pompously presumed. One of the most acclaimed and usually reasonable figures in space science, Alan Sandage, made the following statement:

> The intuitive geometry that is fixed on the senses by that outside spatial frame which gives us our ordinary experience seems Euclidean. Areas increase strictly as r^2, volumes as r^3, using the apparently common-sense definition of r. The concept of spatial curvature is foreign to the intuition and unreal to the non-scientist.[*]

The implication is clearly that those who find it unreal cannot, by someone's definition of the word, be a scientist. Well, I am a scientist to my boots, and I tell you right here and now that space curvature is unreal. Put that in your pipe and smoke it.

The issue of one's interpretation of the data and the definition of measurement—essentially, one's prior choice of method—is absolutely crucial:

> If we observe unmistakeable effects we would say the thing 'causing them' is real ... On this definition space-time curvature is real. The predictions of its *effects* via Einstein's equations are well verified [...] the curvature is measured by the non-Euclidean (parameters). Yet the areas and volumes are *not* measured. What actually is verified is that the formalism of the equations *works* in certain experimental situations.[†]

Emphasis is in the original. The entire exercise boils down to one thing only—that the maths works. It's easy, but abjectly meaningless, to "prove" the existence of parallel universes using bespoke "parallel universe geometry." The vital link between that and measured reality is missing. Please think about this long and hard.

---

[*] Alan Sandage, "Observational Tests of World Models," *Annu. Rev. Astron. Astrophys.* **26** (1988), pp. 561-630.

[†] Alan Sandage *op cit.*

Spatial curvature is not part of physical reality. I can say this with the utmost confidence. Spatial curvature is a mathematical principle, and as such, is indubitably just a mental construct. Whether or not it exists as an independent component of the environment is neither supported nor rejected by the mathematical formalism, and we have no means of checking. There is no reason, logical or otherwise, to suspect that mathematics created and contained in the human psyche exists anywhere outside the cerebrum. "God created the integers," wrote mathematician Leopold Kronecker. "All the rest is the work of Man."

The plane surface underpinning Euclidean geometry is a 2-dimensional construct, in other words, it is *flat*. Straight lines can be drawn upon it, consistent in every way with his definitions and postulates, the angles between them measured, and related to one another in a consistent, systematic way. The surface of a sphere (or part of it), on the other hand, is neither from the inside nor the outside 2-dimensional. The fact that it has an inside and an outside is proof enough. A sphere is 3-dimensional in space, and curvature presented by a sphere occupies length, width, *and* depth. Latitude and longitude plot two of the dimensions, and the curvature itself is the third. It is quite impossible to impose a straight line upon a spherical surface in the Euclidean space we live in. Such a line would be compelled to follow the curvature.

Let's pause here to review the mathematical understanding of space. It's an entirely different way of thinking that challenges our intuition. So, for want of a better description, mathematical "space" is defined by the curvature of a surface, and different geometrical rules were invented to operate it. This is useful to the point of being essential in non-Euclidean geometry, but we shouldn't forget that neither the surface nor its shape is real. The surface doesn't actually exist in physical reality. It's a mental convenience so that we can get our ducks in a row when dealing with abstract ideas. I suppose this is the crux: Can we do without the notion of mentally projected surfaces in analysing our material universe?

I say yes; nearly everyone else says no. You decide.

In this way, the orthodox view of the universe is now that it has a "shape" with an associated geometry—in my view, a very odd idea. It would be analogous to the surface (and *only* the surface) of an expanding sphere. As this balloon is blown up, all points on the skin move apart evenly, but one could travel any distance in any direction and not find an end. This strange visualisation is the *finite-but-unbounded* world of Relativity.* Whether or not we think it an efficient and accurate reflection of physical reality is immaterial. We must accept that despite appearances to the contrary, the Universe should be visualised geometrically as if imprinted upon the surface of what looks to us like a Euclidean sphere, and that, as they say, is that.

I'd like us to dwell a while longer on this analogy of points on the surface of an inflating balloon, which rather cryptically— and illogically—illustrates that while the proposed Universe gets bigger and less dense, objects within it are not actually moving apart. Yes, it gets bigger without increasing radius. There *is* no radius. In their essay "Evidence for the Big Bang," Björn Feuerbach and Ryan Scranton make a stout effort to explain in natural language what all this means. I am afraid though, that as accomplished as they are, the problem of making sense of Big Bang expansion is a bridge too far. In section 1 (b), they point out:

> The word 'expanded' should not be taken to mean that matter flies apart—rather, it refers to the idea that space itself is becoming larger.[†]

Fair enough, I think we get that, even if we don't understand it: The Universe is expanding without the coordinate positions of material objects changing.

Then, in the very next paragraph, they seem to contradict themselves:

> People often have difficulty with the idea that 'space itself expands'. An easier way to understand this concept is to think of it as the distance between any two points in the universe increasing...For example, say we have two points (A and B) which are at fixed coordinate positions. In an expanding universe...the distance between A and B...is always increasing.

What happens to items of matter at coordinate points A and B? What is the difference in principle between "distance always increasing" and "flies apart"? I strongly suggest you visit their

---

* The other often-used analogy—that of raisins in a rising loaf of bread— illustrates well the principle of expansion but fails to satisfy the finite-but-unbounded requirement of the GR universe.

[†] Bjorn Feuerbacher and Ryan Scranton "Evidence for the Big Bang" from the website www.talkorigins.org/faqs/astronomy/bigbang.html .

web page and try to work it out for yourself. At the very least you will then see that I haven't made my point by editing what they say to suit my purpose.

How there could be more space put between objects without their moving apart defies reason. In any case, the expansion of space itself, so easy in differential geometry, was not addressed by any known physics. Foremost among those realists rejecting it was Fred Hoyle. However, it seemed that universal expansion was proven by the rumour of systematic redshift, so Hoyle himself built expansion into his Steady State Cosmology (although it was not the same kind of expansion implied by Big Bang Theory). Hoyle's co-author Geoffrey Burbidge will to this day not hear of objections to cosmological expansion.

Don't feel bad if you fail to grasp the concept of expanding spacetime, and find the whole idea implausible. You're in good company. I don't think anyone can truly claim to understand it, and certainly no one I play cards with has ever convinced me they really get it at all. Even the illuminati themselves seem puzzled. In response to a reader's question in New Scientist, no less a personage than Nobel Laureate Steven Weinberg declared without batting an eyelid, "Good question. The answer is: space does not expand. Cosmologists sometimes talk about expanding space—but they should know better." He was immediately backed up by Astronomer Royal Martin Rees: "Expanding space is a very unhelpful concept." Well, thanks a lot, chaps. We feel better about it now.[*]

Never mind, the show must go on. The next aspect of the Standard Model that we have to get to grips with is how structures in the Universe are organised. Here we must either understand the basics of General Relativity Theory (GRT), or simply recognise the cosmology that emerges from it and accept it for what it is. In order for the solutions of Friedmann to work there was a further very onerous requirement, which may well turn out to be the fly in the ointment. This type of even, symmetrical expansion could only possibly take place if the arena in which it performed was a rather special place. You see, it had to be smooth and regular, and perfectly featureless, in all directions. It is called the Cosmological Principle: The universe in which the Hubble

---

[*] "All you ever wanted to know about the big bang..." *New Scientist*, 17 April 1993, pp. 32-3

expansion takes place is homogeneous and isotropic; it is everywhere the same.

This critically requires that at the scale upon which the universe smoothly expands, there should be no discernable structure, and no boundary condition or "edge" at any distance, in any direction. I quote from Evidence for the Big Bang by Björn Feuerbacher and Ryan Scranton:

> The basic equations for BBT come directly from Einstein's GR equation under two key assumptions: First, that the distribution of matter and energy in the universe is homogeneous and, second, that the distribution is isotropic. A simpler way to put this is that the universe looks the same everywhere and in every direction.[*]

This is the Cosmological Principle, and in assessing Big Bang Theory and universal expansion we should at all times remember that the expansion *cannot work* if the Cosmological Principle does not apply. It is a critical, not arbitrary, parameter of Big Bang Theory (BBT), and has the necessary implication that the Universe has neither boundary limit nor centre. At the scale at which expansion takes place, the Universe should be utterly without irregularity. There would be no juxtaposition of matter and voids, and no variation in specific density.

Well, take a look. We see no end to structure. The entire visible universe consists of material systems with vast spaces between them. There is no need for us to add anything to this. In terms of its own declared parameters, the theory of cosmological expansion just cannot operate in the universe we see around us. That is why those who sired the idea were compelled to supply us at the same time with a universe, forever unseen, with properties so bizarre that it could entertain the cosmological behaviour they so energetically promote. The concept of universal relativistic expansion is simply illogical and completely untenable. It is expansion that isn't really expansion, driven by an explosion that wasn't actually an explosion, into space that isn't there. Apart from that, it makes perfectly good sense. We have thus in these pages taken another step towards defining the particular analytical method that has drained physical science of practically all forward momentum in fundamental discovery. This is a very serious matter indeed, with visible, negative consequences.

---

[*] Björn Feuerbacher and Ryan Scranton, *ibid*.

Amongst the most lucid of astrophysicists studying universal expansion is Yurij Baryshev of the St Petersburg University. He achieved eminence in cosmological circles with the publication of his 2002 book (co-written with Pekka Teerikorpi) *The Discovery of Cosmic Fractals*. In June of 2005 I had the privilege of meeting Yurij at the *First Crisis in Cosmology Conference* (CCC1) in the delightful medieval hamlet of Moncao, Northern Portugal. There he presented a paper entitled "Conceptual Problems of the Standard Cosmological Model," and last year (2008) followed it with "Expanding Space: The Root of Conceptual problems of the Cosmological Physics." I would like to quote here part of the abstract of the latter paper:

> The space expansion physics contains several paradoxes...
> []...Here we present an analysis of the following conceptual problems of the Standard Cosmological Model: the violation of energy conservation...the absence of an upper limit on the receding velocity of galaxies which can be greater than the speed of light, and the presence of a linear Hubble law deeply inside inhomogeneous galaxy distribution. The common cause of these paradoxes is the geometrical description of gravity...[*]

No matter how far we look, we see no end to morphologically distinct structure and no sign whatsoever of a final frontier, or even, for that matter, a horizon of any kind. But then again, there's not an iota of unambiguous observational evidence for a Big Bang, yet it is widely believed; we have in this pervading tenet of human behaviour something as inexplicable, yet as irresistibly seductive, as the exploratory drive in all animal species — a biological, cellular imperative to seek out, then believe with absolute conviction and utter disregard for rational experience, that which is dark and mysterious. Notwithstanding our puzzlement, it is a phenomenon described consistently throughout recorded history; this is nothing new. Every time we have a Standard Model of anything, no matter what, disciples experience an apparently irresistible urge to accept proof *before* the miracle has occurred. It manifests so remarkably in human endeavour that 19th century German theologian Max Müller had to put a name to it: He described it formally in the literature as the Dialogic Process.

---

[*] Yu. V. Baryshev "Expanding Space: The Root of Conceptual problems of the Cosmological Physics" *arxiv*: astro-ph/0810.0153.

But their problems didn't end there. The Big Bang world still wouldn't work, even if the Cosmological Principle were assumed. Observations on opposite sides of the universe showed a remarkable thermodynamic equivalence, indicating an equilibrium that defies Big Bang Theory and challenges the very notion of expansion.

Things at those polar extremes were simply too far apart (100-and-something billion light years, they reckon, after some frenzied calculation) for there to have been communication between them at the speed of light if the Universe has existed for only 13.7 billion years. This should have been a daunting problem, but of course it wasn't. You see, it would appear that scientists developing cosmological theory have awarded themselves the exclusive privilege of being able, at will, to use magic in their equations. They are not constrained by the laws of nature. That's how Inflation Theory came to the rescue of the ailing expansion idea. For reasons that will become obvious, we need to think carefully about Inflation. Let me borrow a definition from Wikipedia:

> In physical cosmology, cosmic inflation is the idea that the nascent universe passed through a phase of exponential expansion that was driven by a negative-pressure vacuum energy density.

The principle employed in developing Inflation Theory comprises one of the main procedural ineptitudes in modern cosmology: An observation is made that is anomalous in terms of the preferred model (apparent isotropic equilibrium across the span of the cosmos), and, rather than change the model, a solution was invented (expansion changed speed) that defied the fundamentals of the very physics upon which the model is based (things moved at greater than the speed of light by a ludicrously high factor), driven by a force invoked for the purpose (oscillating repulsive vacuum energy). I remind you that "vacuum energy density" is assumed in Big Bang Theory, and therefore creation of space includes creation of energy. Although the point has been considered elsewhere in this work, I must emphasise right now that in terms of the model's own axioms, the creation of space clearly violates the principles of conservation of energy and first cause.

In other words, the response to a measurement that flies in the face of a theory is to conjure up a solution based on completely *ad hoc*, thumb-suck physics. In order to justify this latter-day sorcery, the authors retreated behind the impenetrable walls of mathematical sophism. They invoke de Sitter space. De Sitter space, the imaginative creation of revered Leyden mathematician Willem de Sitter in the early 1900s, is the kind that can undergo so-called "exponential metric expansion" and although the distance between points does thereby increase, so does the length of a measuring rod, thus they remain the same number of rods apart. Think about it: *That means that the theory is impervious to the constraints of observation.* These undoubtedly brilliant thinkers treat the Universe as a mere mathematical curiosity rather than the tangibly real world that we live and breathe in every day of our lives.

One question we can ignore no longer is that of infinity. Any half-way decent cosmology must deal with the concept and practice of infinity clearly and decisively, one way or the other. We need at the outset to ask ourselves, "Does infinity exist?" If it does, even only as a mere likelihood, then we need to include it in our philosophy. It is not necessary—indeed, nor is it possible—to be able to conceptualise infinity or see it in a mental picture. All we need to do for the cosmology we are developing is to accept or deny that any such thing exists in the Universe. Is the Universe in *any respect* endless, or is there a concluding frontier, a diaper within which all things come to pass?

Perhaps it would help to put it the other way around. How could infinity possibly *not* exist? My feeling is that there is no need to debate this. Infinity as a parameter of existence is a logical certainty, an axiom of 3-dimensional space. It is the inevitable consequence of causality and conservation. If every effect has a cause, there can have been no absolute beginning. Likewise, if the essence of phenomena is endlessly conserved, the Universe must always remain. Sure, we can avoid contemplating the inevitable continuation of space and time by confining our philosophy to a finite and delineated radius from our point of observation, but that does not exclude the certainty that there is existence beyond our chosen horizon.

One of the most enduring arguments in favour of a finite universe is called Olbers' paradox. No one knows for sure who came up with the idea first, but the person who made the famous description in 1823 was German amateur astronomer Heinrich Olbers. Lord Kelvin published the first formal solution to the problem in 1901, and it has steadfastly remained ever since in the argument basket of those who would prefer a Universe that doesn't go on forever. In a nutshell, Olbers' paradox is this: In an infinite Universe, every line of sight from Earth would sooner or later end up on the surface of a star. Why then, if we are bathed in infinitely accumulated starlight, is the night sky not visually as bright as day?

The key word here might be "visually". Radio astronomy pioneer Grote Reber and classical physicist Paul Marmet showed with data from Reber's innovative ultra-long-wavelength radio telescope in Tasmania that the night sky is indeed bright—we just can't see it with the naked eye. At the 144m wavelength tracked by Reber, the sky has "… high radio intensity all over except for low intensity patches along the Milky Way, with the lowest at the galactic centre."[*] At non-optical wavelengths, it seems there is a uniform, surrounding residue of light from distant stars. This may in addition to resolving Olbers' paradox, also offer a compelling explanation for the CMBR. In Marmet's words,

> At the corresponding wavelength ($\lambda \approx 1mm$), the universe must appear uniformly illuminated as in Heinrich Olbers' model. His paradox no longer exists, since the sky is uniformly bright at that wavelength, as observed by Penzias and Wilson.

If we localise the frame to illustrate how light propagates quantitatively, it becomes quite clear that Olbers' puzzlement was unfounded after all. Our days are "as bright as day" here on Earth because we are so close to the Sun. More local effects by obscuring screens, like heavy cloud and solar eclipses, can vary daytime brightness significantly. The nights on Earth differ quantitatively from day, and can appear pitch-black to the human eye at times. The fullness of the Moon reflects a muted glow back onto the dark side, enough to let us make our way quite easily without artificial illumination. Overall, though, there is a vast difference in light intensity between day and night, between the side of the Earth that faces the Sun and that which lies in the shadow. This is despite the fact that within our home galaxy alone there are bil-

---

[*] Grote Reber, "A Timeless, Boundless, Equilibrium Universe" *Proc. A.S.A.*, **4**: 4 (1982), pp. 482-483.

lions of Sun-like stars shining away for all they're worth, day and night. So why isn't terrestrial night as bright as day? It's very simple; the light sources are just too far away, even those that are right here in our galactic neighbourhood. Extra-galactic stars are that much fainter, and so on, forever.

If light did not diffuse as it went, apparently as the square of distance (the inverse square law), then we would not need telescopes with huge apertures and gigantic mirrors to collect the faded light of distant objects and artificially restore enough of their original brightness so that we can see them. For all sorts of reasons, light intensity fades as it travels, passing resolutely through and around swathes of resistance on its way. It fairly soon becomes negligible—settling it seems to an ambient equilibrium temperature of just 2.7 degrees above Absolute Zero, way too cold to light up the sky, and far beneath the threshold of most terrestrial instruments. It's as if the itinerant radiant energy of the cosmos goes into hibernation, waiting to be swept up again in the genesis of new stars and the rollout of structure. The night sky is not in fact pitch dark; it is just remarkably less well-lit than day because the light sources, though incalculably numerous, are so seriously debilitated by their remoteness.

The signal from stars is simply overwhelmed by distance, and in time, resembles light less than it does darkness. There's nothing sinister about that. It's the way things get done in nature—light dissipates as it goes, and this would appear to be as true for the beam of your flashlight as it is for all the starlight coming from incandescence on the far side of the visible Milky Way population too. Even if there are infinitely many of them, what we are going to see from any point of view is not going to be infinitely bright, so it's quite possible to have an endless array of stars and still have shadows. Perhaps it is as well to take into account that nature does not behave paradoxically. The conundrum is always a facet of our defective interpretation of physical processes, not of the process itself. Olbers' paradox, like all paradoxes perplexing the human mind, does not exist in nature, only in our understanding (or lack thereof) of natural processes.

This somewhat shallow exposition of the problem might be too rich in naivety for most serious scholars. Here is an altogether more sophisticated explanation, for which I acknowledge the assistance of my friend Dave Roscoe of the University of Sheffield:

Olbers' paradox is commonly cited, nowadays, as evidence for a finite universe. But it is not, for it is simple to see at least one way out that is entirely consistent with what is observed: as originally stated, Olbers' paradox assumes that stars are distributed evenly everywhere so that the numbers of stars per unit volume is more-or-less fixed (so long as we use big enough volumes). Put mathematically, this means that the number of stars in sphere of radius R, as observed from Earth, is proportional to the volume of that sphere, and so varies as R^3. One way out of the conundrum is now easily seen: suppose that, instead of the number of stars varying as R^3 (the sphere's volume), the number of stars in the sphere varied as R^2 (the sphere's surface area). This would mean that as we surveyed the stars in ever larger spheres centred on the Earth, the number of stars observed per unit area of any given sphere's surface would remain the same, and so Olbers' paradox just goes away. The very happy fact is that on very large scales (so far as we can tell), the number of stars (well, galaxies anyway) does vary exactly in this way!

Is there anything else in our experience which might indicate a limit to existence? The apparent ceiling in measured redshift values is not a physical frontier. We cannot draw such a conclusion until we understand the cosmological redshift phenomenon in considerably more detail. Even if we were to find what appeared to be some kind of natural, physical boundary condition, we *must* ask the question, "What lies beyond?" It simply cannot, in my humble opinion, be *nothing*. If you're having trouble grasping infinity, try conceptualising absolute nothingness!

It all seems to depend on how we define "universe." The "universe" referred to in Big Bang *must be finite* for the theory to hold. It was at one stage concentrated into a limited amount of space. If all the matter in the universe was packed into this minute sphere, and none left out, then "all" is a finite quantity. If it were infinite, then there would always be some left out, not so? Furthermore, if it has expanded from a finite, concentrated volume, then it cannot now be spatially infinite in extent. It's logically impossible to convert a spatially finite entity into a spatially infinite expanse. It would quite simply never get there. This picture is sometimes countered by suggesting that we are deluded by "thinking Euclidean" in a non-Euclidean world. Hopefully, we're done with that kind of reasoning, but I'm not too confident. For those locked irretrievably into the geometrical approach to physical science, it is probably a bridge too far, and that is a great pity.

The Big Bang expanding Universe just *has* to be spatially finite; it is required by the underlying General Relativity Theory (GRT). But even mainstream theorists doubt it. Eminent Oxford cosmologist Joseph Silk undertook a definitive study to compare geometric models, and controversially declares,

> ...we show that, given current data, the probability that the Universe is spatially *infinite* lies between 67% and 98%, depending on the choice of priors [...] we have shown that a model selection perspective places much more taxing requirements on the accuracy of future datasets than one would naively assume [...] *no amount of observations will be able to decide on the true geometry of the Universe.*[*]

I put the emphasis on both the word "infinite" and the last sentence. I would hope by now that you would too. Bluntly put, the world's favourite cosmological model is extremely unlikely, so much so that it is astonishing that it should have received more than just fleeting curiosity from astronomers. With hindsight, we can see what happened: We were ambushed!

Unfortunately, cosmology in the 20th century was considered to be confined to one of only two camps—the expanding fireball hypothesis claimed by George Gamow and the Steady State theory promoted by Fred Hoyle. It was in comparing these two models (with wilful bias to the former, it must be said) that we arrived at the appalling justification for the methodology of modern, consensus cosmology: Science as the lesser of two evils. Many times we hear someone say, in all earnestness, "We accept Big Bang Theory although it leaks like a sieve, simply because it seems to work better than the alternatives."

That's science? Sounds like a game of gin rummy to me.

In cosmology, we would be forgiven for thinking that preference is for the *greater* evil, since Steady State generally employs far better physics than Big Bang Theory (and please, I imply no literal sense to the word "evil"; neither theory is wicked, merely imperfect to some degree). Success in the arena of competing ideas has always boiled down, not to bricks-and-mortar science, but to the measure of charisma and evangelical skill of the pro-

---

[*] Mihran Vardanyan, Roberto Trotta, and Joe Silk "How flat can you get? A model comparison perspective on the curvature of the Universe" *arxiv*: astro-ph/0901.3354.

ponents. George Gamow was a star on the stage and Fred Hoyle only marginally less so. Hoyle saw it clearly:

> More and more the professions will cross over into the entertainment field. Those of us who are not employed directly in industry will come to realize that what we are really in is 'show biz.' [*]

The others, I'm afraid, were rather dull. Consequently, many fine ideas fell by the wayside, including, in my opinion, the most realistic model of all: The static, equilibrium, endless Universe first proposed in 1928 by Nobel Laureate and author of the 3rd law of thermodynamics, Walther Nernst.

Nernst's conception of a stationary Universe was supported and developed by illustrious physicists of the 20th century, including Erwin Finlay-Freundlich, Max Born, and Louis de Broglie; two of whom would themselves subsequently receive the Nobel Prize. Despite being crafted with exemplary physics based upon observation, it has been studiously ignored by mainstream space scientists for more than 80 years. Why cosmology should have been limited to a two-horse race—and latterly, to a one horse race—when there is so much good science opposing the sole remaining nag on the course is quite beyond me.

It would be sensible now to consolidate the principles discussed so far: We have established by being carefully retrospective that the mathematical properties of space are not by their articulation made externally real. Geometry remains simply a basis for the measurement of material objects and the relationship between them. The transmutation of geometrical concepts from objects *in* space to space itself is a slipped disk in Mathematics' skeleton, and we bear the brunt of that aberration to this day.

With the foregoing in mind, it is time to ask ourselves precisely *why*, in the clear absence of visual cues, we first believed that the Universe is expanding, if indeed we do. This is the most important question that can be addressed to any article of faith or intransigently held opinion—can one's motivation be linked at all to the comfort of subservience and fear of authority, though we might well know it's wrong? Or worse still, to the preening flattery of oneself that something is real just because we can imagine it?

---

[*] Fred Hoyle *Of Men and Galaxies* (New York: Prometheus Books; 2nd edition, 2005).

Recently, Paul Jackson, a patient friend and guiding light in my journey towards greater understanding, sent me a tailpiece to an earlier conversation we'd had on whether it is sensible to say that some parts of Black Holes, for example the event horizon, are physically realisable. Could we in principle travel in a space ship towards a suspected Black Hole candidate and bump into such a thing? It turned out to be a long and exhausting dialogue in which we struggled to find some logically satisfying answer to what is really a very simple question. We'd agreed that we should not let our enquiry sink to the level of technical discussion on the syntax of pure mathematics. We need to examine the principles conceptually, describe them in natural language, and see if they make sense. Paul suggested I look at a paper by Canadian theorist Ivan Booth, in which he declares,

> It is certainly possible that while event horizons may not be particularly physical there may also be no replacement for them. In this case black holes would be fundamentally different from other objects in the universe such as stars, elephants, and toasters. [*]

It turned out to be a very useful discussion after all. I'm still trying to distil the essence of my approach to things like Black Holes and Big Bangs—an approach my mentor, the late Tony Bray once called "agricultural astrophysics." He was not being unkind; in fact, I believe it is a backhanded compliment. When I looked at Booth's paper (I didn't get very far before I was tangled in the web) a glimmer of understanding of my own attitude flickered in the darkness. The question I ask myself is, "Why do we do this?" Surely what has been observed is enough to occupy us as astrophysicists? Why on Earth do we put ourselves through the mental agony of trying to decipher the nature of Black Holes? Maybe there's a deeper psychological reason. Perhaps we take existential (dare I say *perverse*?) pleasure in the mental game, even if it has no foreseeable happy ending. There is a nearly irresistible compulsion in all of us, as far as I can make out, to pursue dark and mysterious stuff that may or may not exist when all the while there's an elephant in the room. I'm trying very hard to take a different route in life.

---

[*] Ivan Booth "Black hole boundaries" *arxiv*: gr-qc/0508107.

I related in my book *The Virtue of Heresy*[*] an account of a memorable supper I was privileged to attend some years ago. I shared a table with Huseyin Yilmaz, Caroll Alley, and Hal Puthoff. The latter two are experimentalists and teachers, professors of physics, while Yilmaz, then well into his 90s, is a theorist who worked with Einstein and Wheeler at the Institute for Advanced Studies, Princeton University. There I sat with wide eyes and dropped jaw as these gentlemen discussed General Relativity. Back and forth went the arguments, led always by "Yes, but...": GR does not conserve energy, only energy-momentum; r is not a radius; yes, r is a radius; the Schwarzschild solution does not give singularities or event horizons; yes it does; GR *excludes* gravitational stress-energy as a source of curvature; stress-energy is merely a coordinate artefact in GR, a "pseudo-tensor, which can appear or disappear depending on how you treat mass." And so on. The debate went on for about 3 hours, until Professor Yilmaz fell asleep. I thought to myself, "Heck, this argument took place between blokes who actually agree with each other!" I became convinced right then that there just *has* to be a better way to do science. Surely?

So I decided to try to do it by dealing only with "stars, elephants, and toasters," as Booth put it, and that meant eschewing ghostly science and superstition. If some poor soul had a dream last night that the world was being swallowed by a rabid Calabi-Yau manifold in dodecahedral Poincaré space, good luck to him. I hope he gets over it. I can't help; I'm too busy figuring out the apple in my hand, and the sky density of quasars, and the chemical composition of the Sun, and a world that accommodates tyrants, to be able to give serious thought to a chap who says my apple's actually a frog.

The single most important idea I have sought to convey to you in these pages is this: Mathematics has risen from its practical roots to an exalted place where it dictates what we *should* see. We do not *see* the spacetime, curved or otherwise, dictated in General Relativity; we cannot *see* universal expansion imposed upon us as a result of that theory; yet we are compelled from every side to visualise the world that way. We are instructed to create this image in our minds, and so, in a tragicomic parody of the scientific

---

[*] Hilton Ratcliffe *The Virtue of Heresy—Confessions of a Dissident Astronomer* (Charleston: BookSurge, 2008).

method, we humble ourselves under a barrage of intellectual gobbledegook to create the desired picture of the Universe with our eyes wide shut.

It's time, ladies and gentlemen. Let's tot up, see what we owe, and settle the bill.

- Evidence in favour of expansion: Optimistically, about half the equations, no observations whatsoever, and at least three quarters of current propaganda.
- Evidence against universal expansion: *All* observations, and the remaining half of the equations.

Quod erat demonstrandum!

> **Discussion**: *The shape of space and its ability to erupt are nothing more than mind games, quite incapable of being verified observationally. Expansion is simply a mathematical construct emerging from one of three solutions to a particular set of equations, and the choice is arbitrary.*

Chapter 9

Consensus Science

All those in favour, say "Aye".

A fireside précis of preceding arguments: We have no good reason to sustain belief in universal expansion while observations at all sides tell us otherwise. "There are things we know, things we know we don't know, and then there are things we don't know we don't know." (Glenn Starkman, "Introducing doubt in Bayesian model comparison," 2008).

◆ ◆ ◆ ◆ ◆ ◆ ❖ ◆ ◆ ◆ ◆ ◆ ◆

*A*t the outset, it was clear that no certainty could be reached in cosmology. We are dealing with no more than likelihood and probability, and are best-guessing a picture of the ultra large world beyond our observational horizon. The challenge this book faces in unravelling the plot is clearly focussed on dismantling the psychological obsession that drives a model built upon whispers. We are not so much questioning *what* people believe as pointing out that whatever it is, it did not emerge from rational debate. We simply enquire, "Why do you believe what you do, and how is it that your faith is held with such certainty that you guarantee the correctness of your own opinions by creating a titanic theoretical edifice based upon them?"

However, all is not lost. Not everyone is totally happy with the direction cosmology has taken. Two modern astronomers who obviously find room for improvement in the current consensus model are Vicent Martinez and Virginia Trimble:

> It seems clear that many of the pre-Copernican astronomers who made earth-centred models gradually more complex to match better observations thought—according to historians, anyhow—that they were describing the phenomena, not explaining them. Are cosmologists continuously re-editing an undeclared unsuccessful model of the universe to accommodate it to new and unexpected observations? Several are already declaring the crisis of the present cosmological model and advocating for the need of a paradigm shift in a Khunian sense but, at the same time, the general adherence to the mainstream concordance Lambda-CDM model does not leave too much room for thinkers outside the accepted cosmic paradigm. Does this mean that theorists or observers or both should give up on the universe and go back to studying cataclysmic variables (of which we are secretly very fond)? Certainly not![*]

In the cold light of day, it is mystifying that an entire philosophy of the Universe, something approaching a Theory of Everything, could have been constructed in fragile layers, each in turn demanding that we blindly accept ideas that have no unambiguous analogue in the world around us. Any theory needs to be tested against external reality, and the way to do it is by objective empiricism and the classical scientific method. Some might doubt this. There are many in the world of contemporary meta-science who argue that we are ill-equipped to logically contemplate a common material reality, and indeed, in some extreme cases, insist that there is no such thing.

If we proceed from the standpoint that human perceptive abilities cannot give us meaningful answers because they are flawed, or at the very least have obvious limitations, we are hoist by our own petard. We are thereby clearly prevented from deriving scientific truth about the mysteries of our environment, all of us, and should in consequence all simply give up and go home for tea. We would condemn ourselves to dumb, existential solitude or to a social club where we blissfully preach to the converted.

> We believe the most charitable thing that can be said of such statements is that they are naive in the extreme and

---

[*] Vicent Martinez and Virginia Trimble "Cosmologists in the dark" *arxiv*: astro-pf/0904.1126.

Figure 35: Participants in the groundbreaking First Crisis in Cosmology Conference, Portugal, 2005. The author is centre front, with striped shirt and clasped hands. On my immediate left (to the right of me in the picture) is Oliver Manuel, discoverer of neutron repulsion. He is still denied the Nobel Prize for the most fundamentally important physical discovery of the 20th century.

betray a complete lack of understanding of history, of the huge difference between an observational and an experimental science, and of the peculiar limitations of cosmology as a scientific discipline. By building up expectations that cannot be realised, such statements do a disservice not only to astronomy and to particle physics but they could ultimately do harm to the wider respect in which the whole scientific approach is held. As such, they must not go unchallenged. It is very questionable whether the study of any phenomenon that is not repeatable can call itself a science at all. It would be sad however to abandon the whole fascinating area to the priesthood. But if we are going to lend this unique subject any kind of scientific respectability we have to look at all its claims with a great circumspection and listen to its proponents with even greater scepticism than is usually necessary. This is particularly true when the gulf between observers and theoreticians is as wide as it usually is here.[*]

Is the Universe expanding? It would appear not. What do we see? We do not *see*, let alone measure, large objects systematically moving away from all other large objects. On the contrary, it would seem to be quite the opposite, at least in the case of colliding spiral galaxies. Every observable large scale system is to all

---

[*] Mike Disney, "The Case Against Cosmology" *arxiv*: astro-ph/0009020.

Figure 36: The Whirlpool Galaxy, M51. The spiral follows the classic Fibonacci curve found ubiquitously in nature. (Image courtesy of NASA, STScI, and Hubble Heritage).

intents and purposes in a state of equilibrium, even if it might be expressed dynamically as a cycle. Is the Universe in any sense, on any axis, *finite?* It might be, in theory anyway, but where is the evidence? Of course, we cannot observe anything infinite; but then again, neither do we detect even the faintest sign that the Universe reaches finality. We do not come across any kind of absolute boundary condition.

In terms of standard physics, Hubble expansion related to cosmic redshift has failed, even after several restarts. We should not even bother trying to explain to ourselves and interested spectators the functions of expanding space, or how the rate of expansion varies at the convenience of our Standard Model. We need go no further than simply examining the pictures we get of the sky. Our observations show, with as much certainty as can be expected over cosmological distances, that the expected direct association of higher redshift with a more immature Universe has not materialised. Modelling the universe in onion-skin layers of redshift values fails dismally to show with greater redshift the least sign of higher density; smaller object size; higher temperature; lower metallicity (or higher metallicity); smaller voids; less apparent *and* intrinsic brightness; infant galaxies; or any other sign that redshift truly indicates remoteness and youth in an evolving Universe. To make matters truly embarrassing for the Standard Model of Cosmology, the redshift patterns that supposedly indicate and verify an expanding cosmos have been found in local space, well within the confines of Virgo. We all know that by consensus, theory excludes local space from expansion, so the signs in the sky are Judas goats, leading us to the nemesis of redshift-based cosmology.

In addition, given that the redshift of Hubble expansion goes hand-in-hand with the Cosmological Principle and cannot exist without it, it is of crucial importance that we note no smoothing out of the cosmos, no matter how high the redshift value. There is structure, great big lumps of it, for as far as we can see. The redshift distance ladder is obviously flawed, and with it our 3-D conception of cosmic geography. To top it all, some well presented observations show that there are objects that deny their redshift-given remoteness by the fact that their transverse expansion would then exceed the speed of light, many times over. That alone crushes the concept without hope of redemption. The redshift-expansion idea, despite the concerted efforts of the finest scientists on Earth to promote it, has failed.

What of the Cosmic Microwave Background Radiation? Is it really a picture of a very dense, nascent Universe? I doubt it, and by now, I should hope that you doubt it too. The controversial "predictions" of BBT concerning an enveloping primeval radiation signature are hopelessly lacking in true predictive power, and in addition, quantitatively way off beam. Alignments with local astrophysical structure (and voids) are routinely confirmed by WMAP analysts. The CMBR was from the word go a hopelessly optimistic long shot. Analysts are kept busy, night and day, trying to cope with anomalies—that is, disagreements between the image and what is expected by the model. They have, despite great effort and inventiveness, thus far failed abysmally to get that obstinate, hee-hawing picture to fit the theory.

How we react to anomalous results is going to be crucial to the future of cosmology, the empirical foundation of astrophysics, and indeed, possibly the importance of scientists to the progress of society generally. The sincerity with which we incorporate discordant results into our knowledge base and theoretical structures will in my view define the relationship between astronomy and cosmology, and may well determine whether such a link can exist at all. The anomalies result always, and exclusively, from our comparison of the data with theoretical models. The data and images are not in and of themselves anomalous, and cannot be intrinsically peculiar. Neta Bahcall puts it rather well:

> The advantages of 'What you see is what you get' ...may
> be more important than the elegance of the solution.[*]

Whether we continue to pursue the mysteries of the larger-scale cosmos with our eyes wide shut, or instead with due circumspection take notice of the measurable reality surrounding us, time will tell.

Let's be honest. We are unable to measure the global physical divergence of galaxies. There is no unambiguous, empirically tested correlation of redshift velocity with distance. We cannot observationally verify a proposed universal geometry that would permit expansion. No deep sky survey has revealed evolution with time in astrophysical objects. An image of the primordial fireball (or any other deity) can only be seen in background radiation by express construction, and even then, through the rosiest of rose tinted spectacles. These things are best described as superstition, but we make no judgement on people who are superstitious; we merely try our damndest to separate them from rational science.

Our uncertainty is admittedly less with nearby things, but the incredible vastness of our field of study is such that even within the Solar System itself, we are unsure of most things. We don't know with any clarity how big the Solar System is; what it consists of; or what keeps it going. Where did it come from, how did it form, and whither next? The same is true for the Local Group of galaxies. The caveat remains that they certainly show no measurable sign of the creation of spacetime within their boundaries. We must accept expansion with only the reputation of our forefathers to go on.

> Thus it must be remembered that the whole argument is
> based on the idea that helium was made by such a fireball,
> and much as most people want to believe it there is no independent
> evidence that this ever did take place. Most of

---

[*] Neta Bahcall, "Large Scale Structure in the Universe Indicated by Galaxy Clusters" *Ann. Rev. Astron. Astrophys.* **26** (1988), pp. 631-686.

the helium was made in a big bang, and the parameters required are those chosen in the conventional model. This is the most popular view but in its present form it requires that we choose an initial photon/baryon ratio, invoke a 'magical' inflation era, and assume the presence of a large amount of dark nonbaryonic matter, and dark energy (creation energy). These are four assumptions for which we have no basic theory, nor direct observational evidence. Just authoritarian belief.[*]

So, when all is said and done, it comes down to this: Is what we see relevant to the formulation of cosmology? To sustain a well-worn cliché—do we believe what we see, or see what we believe? It's a choice really; one which will determine whether the *status quo* remains and dictates reality, or whether we do indeed live at the cusp of revolution. I am under no illusion; the odds against my preferred outcome are almost impossibly huge, but a light at the end of my tunnel is kept flickering by the knowledge that it has happened before, time and again. This regime must fall, that is certain, but when? Perhaps you, the few, will determine by what you do next what the outcome shall be.

I shall leave you with a quote from the essay "Modern Cosmology, Science or Folk Tale" by Mike Disney:

> It may be healthier, as well as more exciting, to admit we are surrounded by great mysteries which will provide challenges for generations to come. More fundamentally, as Daniel Boorstin the historian of science remarked: 'The great obstacle to discovering the shape of the Earth, the continents and the oceans was not ignorance but the illusion of knowledge. Imagination drew in bold strokes, instantly serving hopes and fears, while knowledge advanced by slow increments and contradictory witnesses.' If we are not appropriately sceptical about cosmology today then the current myth, if myth it is, could likewise hold up progress across all of extragalactic research for generations to come.

Ultimately, perhaps, we have attempted to address in our book a single question, one that is supremely difficult to answer with conviction: *Is the Universe expanding?* We are baffled for one simple reason—by definition, the expansion described in the Standard Model of Cosmology occurs exclusively beyond the

---

[*] Geoffrey Burbidge, "B²FH, the CMB, and Cosmology" *arxiv*: astro-ph/0806.1065.

reach of measurement. If for no other reason than that, such a supposition should be excluded from the realm of reasonable science.

Thank you for sharing this journey with me, and good luck to you.

Rules of Reasoning in Philosophy

Rule I

We are to admit no more causes of natural things than such as are both true and sufficient to explain their appearances.

Rule II

Therefore to the same natural effects we must, as far as possible, assign the same causes.

Rule III

The qualities of bodies, which admit neither intension nor remission of degrees, and which are found to belong to all bodies within reach of our experiments, are to be esteemed the universal qualities of all bodies whatsoever.

Rule IV

In experimental philosophy we are to look upon propositions collected by general induction from phenomena as accurately or very nearly true, notwithstanding any contrary hypotheses that may be imagined, till such time as other phenomena occur, by which they may either be made more accurate, or liable to exceptions

Taken from Isaac Newton's *Philosophiae Naturalis Principia Mathematica* Translation: Cohen, I. Bernard, and Whitman, Anne. Berkeley: University of California Press, 1999.

Addendum 2

Classical Conventions

Latin (Roman) Prefixes of Scale

Multiple	Prefix	Symbol	Multiple	Prefix	Symbol
10^{-1}	deci	d	10^{1}	deca	Da
10^{-2}	centi	c	10^{2}	hecto	H
10^{-3}	milli	m	10^{3}	kilo	K
10^{-6}	micro	μ	10^{6}	mega	M
10^{-9}	nano	η	10^{9}	giga	G
10^{-12}	pico	p	10^{12}	tera	T
10^{-15}	femto	f	10^{15}	peta	P

Commonly Used Greek Letters:

Capital	lower	Name	Usage (examples)
A	α	alpha	
B	β	beta	
Γ	γ	gamma	Gamma rays
Δ	δ	delta	Change, variation
H	η	eta	
Θ	θ	theta	
Λ	λ	lambda	Wavelength; Cosmological Constant (Ω_Λ)
M	μ	mu	
N	ν	nu	Velocity; neutrinos
Π	π	pi	
P	ρ	rho	Pressure; density
Σ	σ	sigma	Frequency
T	τ	tau	
Y	υ	upsilon	
Ψ	φ	psi	Wave function
Ω	ω	omega	Density

Addendum 3

Abbreviations

2dF GRS - 2dF Galaxy Redshift Survey
2Q2 - 2Df Quasar Redshift Survey
ACG - Alternative Cosmology Group
AGN - Active Galactic Nucleus
ApJ - Astrophysical Journal
ASSA – Astronomical Society of Southern Africa.
BBT - Big Bang Theory
BH - Black Hole
BSO - Blue Stellar Object
CCC1 - 1st Crisis in Cosmology Conference
CCD - Charge Coupled Device
CERN - Centre for Research into Nuclear Energy (French)
CfA - Centre for Astrophysics
CMBR - Cosmic Microwave Background Radiation
CME – Coronal Mass Ejection
COBE - Cosmic Background Explorer
CREIL - Coherent Raman Effect in Incoherent Light
EMR - Electro Magnetic Radiation (*i.e.* light)
ESA - European Space Agency
ESO - European Southern Observatory
FLWR - Friedmann-Lemaître-Walker-Robertson (metric)
FRS - Fellow of the Royal Society
FRW - Friedmann-Robertson-Walker (metric)
GLY - Giga Light Years (1 billion LY)
GRB – Gamma Ray Burst (or Burster)
GRT - General Relativity Theory
HCU - Human Consciousness Unit
HST - Hubble Space Telescope

HUDF - Hubble Ultra Deep Field
IGM - Inter Galactic Medium
ISM - Inter Stellar Medium
L-C & G - López-Corredoira and Gutiérrez (astronomers)
LCDMM - Lambda Cold Dark Matter Model. Also ΛCDMM
  or λCDMM.
LHC - Large Hadron Collider
MLY - Mega Light Years (million LY)
Mpc - Mega Parsec (million parsecs)
MPE - Max Planck Institute for Extraterrestrial Physics
NGC - New General Catalogue
QCD - Quantum Chromo Dynamics
QED - Quantum Electro Dynamics
QSO - Quasi Stellar Object (quasar)
QSSM - Quasi Steady State Model
SALT - Southern African Large Telescope
SDSS - Sloan Digital Sky Survey
SKA – Square Kilometre Array radio telescope.
SMBH - Super Massive Black Hole
SN - Supernova
SNe - Supernovae
SNO - Sudbury Neutrino Observatory
SRT - Special Relativity Theory
TFR - Tully Fisher Relationship
VLT - Very Large Telescope
VWL - Visible Wavelength Light
WMAP - Wilkinson Microwave Anisotropy Probe

Glossary

(In some cases, definitions are given both in plain English and in the language of physics).

- **Aberration of starlight.** Apparent displacement of the star from its true position caused by the combined effect of the speed of light and the speed of the Earth around the Sun (30km/sec).
- **Acceleration.** A change in the rate (higher or lower) or direction of motion.
- **Active galaxian nuclei (AGN).** The core areas of galaxies in turmoil.
- **Adiabatic.** A process without transfer of heat.
- **Analogue.** Continuous, unbroken stream, as in information displayed on a dial.
- **Anisotropy.** The manifestation of different characteristics when measured in opposing directions along an axis; uneven, asymmetrical distribution. (See: **Isotropic**).
- **Antimatter.** The inverse value of *matter*.
- **Array.** *Mathematics*: Symbols arranged in columns and rows. *Astronomy*: Collection devices linked together to increase their power and resolution.
- **Astronomical unit (au).** The average distance of the Earth from the Sun, approximately 150 million kilometres.
- **Atom.** The smallest unit of matter; atoms are the fundamental components of a chemical reaction; a combination of protons, neutrons, and electrons.
- **Atomic number.** Symbol Z, the number of protons (positive charges) in the nucleus of an atom.
- **Baryonic matter.** Matter comprised of protons, neutrons and electrons; the *standard model* for matter.
- **Big Bang.** A hypothetical event, resembling a hybrid of an explosion and an implosion, postulated to mark the origin of the universe and the beginning of time; estimated to have occurred 13.7 billion +/- 800 thousand years ago.

- **Billion**. One thousand million, 1,000,000,000; also expressed 10^9.
- **Binary**. An expression using only two distinct elements. Binary notation is a system of numbers using only zeroes and ones.
- **Binding energy** (aka *nuclear binding energy*). The difference between the measured mass of an atomic nucleus and the total of the individual masses of its constituent parts. Technically, the energy released if a nucleus of atomic number Z and mass number A is made by combining Z atoms of H-1 with (A – Z) neutrons.
- **Blackbody**. An idealized, theoretical surface that absorbs and emits all radiation incident upon it, and is therefore in thermal equilibrium. It has no capacity for reflection. Stars are assumed to be *blackbodies* for purposes of describing stellar radiation.
- **Blackbody radiation**. Thermal radiation from a blackbody, which follows a characteristic curve of energy and temperature for any given wavelength; displays a *Planck spectrum*.
- **Black Hole**. Hypothetical concentration of mass such that escape velocity exceeds the speed of light; controversial theoretical construct arising from arbitrary solutions to General Relativity.
- **Bolometric**. Involving radiant energy.
- **Brownian motion**. The chaotic motion of molecules in a gas.
- **Cepheid variables**. Stars with regularly fluctuating radiant output, used as "milestones" for the calculation of astronomical distances.
- **Celestial sphere**. The imagined inverted dome upon which the characteristic stellar patterns appear.
- **Centre of gravity**. A geometrical point in any material system at which the gravitational potential of the system is directed.
- **Cepheid variables**. A class of stars of oscillating intensity used as distance indicators.
- **Chaos theory**. Variations in the outcome of events that, although subject to deterministic laws, are nevertheless influenced by ambient variables, *e.g.* the growth of a snowflake, weather forecasting.

- **Cherenkov light**. A luminous flash indicating the occurrence and direction of a neutrino event.
- **Chromosome**. Rod-like minute structure present in cell nuclei during division; contains *genes* and transmits hereditary characteristics.
- **Cluster**. A system comprised of individual parts grouped close together; of stars and galaxies, coherently bound actual (not apparent) groupings of such entities.
- **Constant flux**. A cosmological theory describing a continuous sequence of universal expansion and contraction.
- **Conservation of energy (matter)**. The axiom that the existence of something precludes the possibility of nothing.
- **Cosmic Microwave Background Radiation (CMBR)**. Roughly isotropic short-wave radio noise enveloping the Earth, asserted to be an image of the primordial fireball.
- **Cosmogony**. Study of the evolution of the Universe as a whole or of a component system within it
- **Cosmography**. Universal equivalent of *geography*.
- **Cosmology**. That part of astronomy that seeks to describe the origin, evolution of the Universe; in its current incarnation, pseudo-scientific religionism.
- **Cosmos**. The Universe seen as a disciplined system.
- **Cosmological**. Property or effect of the (nonlocal) *Cosmos*.
- **Cosmological constant**. A mathematical term (temporarily) introduced to General Relativity to suppress the Friedmann solution indicating expansion or contraction.
- **Cosmological principle**. The hypothesis attributed to E. A. Milne, that the large-scale Universe is homogeneous and isotropic.
- **Coulomb energy**. Repulsive energy between the charges of protons; reduces *binding energy*.
- **Dark energy**. An imagined repulsive force introduced to account for the acceleration of universal expansion.
- **Dark matter**. An imagined attractive force introduced to account for perceived mass anomalies in astrophysical systems.
- **Digital**. Divided into units. In mathematics, usually a sequence of numbers. Used here as a synonym for *quantised*.
- **Dogma**. Philosophy held intransigently. Unreasonable standpoint.

- **Doppler effect.** A change in radiated wavelength due to relative motion.
- **Dualism.** A theological term describing the separation of body and soul, man and God.
- **Duality**. Regressional equivalent of singularity; the polarised form in which reality presents itself.
- **Ejecta.** Material expelled from a progenitor.
- **Electricity.** Stream of electrons in a conducting medium.
- **Electromagnetic force.** The cohesion of radiated energy, one of four fundamental forces of nature.
- **Electromagnetic radiation**. Light at all wavelengths, visible and extra-visible.
- **Electron**. Negatively charged particle found outside of the nucleus in all un-ionised atoms.
- **Energy**. The capacity to do work and overcome resistance; the manifestation of that potential.
- **Entropy**. Increased complexity.
- **Energy Parity Level**. A momentary balance between mass and kinetic energy; temporary tidal equilibrium. See *Parity*.
- **Escape velocity**. The minimum velocity that will allow an object to overcome gravity. Approximately 40 000 km/h on Earth.
- **Event**. An interaction of energy that may be encapsulated by space-time coordinates.
- **Event Horizon**. The surface of an imaginary sphere marking the boundary of a black hole.
- **Event Threshold**. A point in space-time marking the commencing point of evolution, before and beneath which no further simplification can occur.
- **Evolution.** The progressive, open-ended transformation of systems with time, usually selectively driven by function.
- **Field.** The spatial arrangement of energy potential.
- **Force.** A condition with the potential to rearrange matter, or change its rate of motion; a dynamic influence on acceleration.
- **Foreshortening**. Aka *Lorentz-Fitzgerald contraction*. A balancing factor introduced to Special Relativity to allow an absolutely constant speed of light.

- **Galactic**. Adjective referring specifically to components and properties of the Milky Way (also referred to as *The Galaxy*).
- **Galaxian**. Adjective referring specifically to components and properties of galaxies *other than* the Milky Way.
- **Galaxy**. A large collection of stars held in a system by an integrating mass energy. More accurately: *A vast formation of plasma clouds that contain electrical currents and occasional, widely distributed tiny lumped points of matter called nebulae, stars, and planets.* (Latter definition given by Donald Scott).
- **Geocentric**. Earth-centred (in cosmology).
- **Geodesic**. Path of least resistance followed by matter in Einstein's curved space-time.
- **Gravitational lensing**. The bending of light by a foreground gravitational field to create an image of a background object.
- **Graviton**. Postulated particle that is the carrier of gravitational force.
- **Gravity**. The force of attraction between objects with mass. One of four fundamental forces of nature. Considered by *Relativity* to be an effect caused by curved space-time.
- **Hadron**. In Particle Theory, a class of elementary particles comprised of still smaller particles called *quarks* and *antiquarks*.
- **Half-life**. The uniform amount of time taken for half of a given unstable isotope sample to decay.
- **Hawking radiation**. Particles that escape from a black hole.
- **Heat**. The transfer of energy from a body with a higher temperature to a body with a lower temperature. This is work at a molecular level without the presence of external forces.
- **Heavy water**. Water formed partially with deuterium.
- **Heisenberg's uncertainty principle**. See *Uncertainty Principle*.
- **Heliosheath**. Suggested boundary to the Solar System.
- **Hubble constant (H_0)**. The rate at which the Universe is said to be expanding.
- **Hubble law**. Proportionality seen between cosmic redshift and recessional motion in expanding universe theories.

- **Human Consciousness Unit (Shell).** The finite range of phenomena being investigated in this book
- **Ideological Momentum:** The impetus of collective opinion; the tendency for supportive results to emerge from prior consensus or authority; also called 'the snowball effect'; a synthetic trend in which we impute meaning in things just because we *want* meaning to be there for whatever deeply held reason, and then take that meaning forward even when it has been objectively falsified.
- **Inertia.** Resistance to acceleration.
- **Infinity.** A state of being limitless, unending.
- **Inflation.** Early period of extreme superluminal expansion invoked to enable Big Bang theory.
- **Intelligence.** The ability to rationalise *and* appreciate aesthetics.
- **Interferometer.** An instrument of extremely accurate small-scale measurement (wavelengths, angles) using the principle of *interference fringes* in light.
- **Investment bias:** Influence on results stemming from the need to satisfy sponsors.
- **Ion.** An atom having electric charge, *i.e.* unequal numbers of protons and electrons.
- **Isotope.** Atoms having the same atomic number but different atomic weights (different numbers of neutrons in their nuclei); nearly identical chemically, but physically different.
- **Isotropic.** Having properties that do not vary with direction.
- **Jets.** In astrophysics, a collimated outflow of matter.
- **Kinetic energy.** The energy of motion, *e.g.* momentum.
- **Light.** Commonly, visible part of the range of electromagnetic radiation. Some light is invisible to humans.
- **Light year.** The distance that light would travel in a vacuum in one year: 5 trillion 869 billion 713 million 600 thousand miles, or ∼9.3 trillion kilometres.
- **Luminosity.** In astrophysics, the intensity with which a celestial object shines.
- **Mass.** The amount of matter an object contains, and therefore its resistance to force; a quantity of matter, defined by two properties, *inertia* and *gravity*.

- **Mass energy**. The energy of attraction between systems, *e.g.* magnetism, gravity.
- **Mass number** *(aka atomic mass number)*. The sum of the numbers of protons and neutrons in an atomic nucleus. This is identical to the atomic mass expressed as a whole number in terms of *atomic mass units* (AMU).
- **Matter**. A form of energy that has substance, inertia, and coherence.
- **Million**. One thousand times one thousand; 1,000,000; 10^6.
- **Model**. Scientific: A mathematical approximation of a supposedly real situation.
- **Molecule**. A compound of two or more atoms held together by a chemical bond.
- **Momentum**. Inertia in motion.
- **Nucleus** (of an atom). Central component of an atom, consisting of proton(s) and neutron(s) bound together by the *strong nuclear force*. Note: The nucleus of a hydrogen atom consists of a single proton only, and no neutrons.
- **Nucleosynthesis.** The formation of atomic nuclei by the binding of protons and neutrons in high-temperature plasmas.
- **Neutrino**. A particle, occurring as a by-product of nuclear fusion, with no mass and no electrical charge.
- **Neutron**. An electrically neutral elementary particle slightly more massive than a proton or atom of H-1; component of atomic nuclei.
- **Neutron star**. The postulated remains of a star that has suffered gravitational collapse, with the resulting preponderance of neutrons at its core. *Pulsars* are believed to be rotating neutron stars.
- **Nova**. Exploding star. See *supernova*.
- **Nuclide**. Name given to an atomic assemblage of a specific number of electrons, protons, and neutrons, *e.g.* H-1, He-4, Fe-56, U-238.
- **Observer effect**. Property of an observer that causes it to perceive an event subjectively.
- **Occam's razor** (aka **Ockham's razor,** due to 14th century English friar William of Ockham). In an hypothesis, making no superfluous argument; striving for the simplest effective solution.

- **Olbers' paradox:** The question "why is the sky dark at night?" is usually taken rhetorically to imply a finite material universe. Other explanations are that light degrades with time and space to become invisible, or that we are shield from background radiation by foreground material.
- **Orbital velocity.** The speed at which a satellite must travel in order to maintain orbit for a given altitude.
- **Parity.** In classical physics, parity is *space-reflection symmetry*, which holds that in all phenomena described by classical physics, no distinction can be made between *left* and *right*. Not to be confused with Energy Parity Levels.
- **Parsec.** A unit of length used in astronomy for extra-big distances. Derived from the properties of an isosceles triangle with a base of one astronomical unit, it is equal to +/- 3.26 light years, although it is usually used as *Megaparsecs*.
- **Perception.** The ability of sentient beings to translate sensory input.
- **Periodic table (of elements).** Systematic list of chemical elements arranged in order of atomic number.
- **Phase transition.** Change in the form of matter from solid to liquid to gas, as ambient conditions change.
- **Philosophy.** An academic discipline concerned with the clarification of the significance to mankind of natural phenomena.
- **Photoelectric effect.** The generation of an electric current by certain substances when exposed to light.
- **Photoionization.** The stripping of electrons from atoms by exposure to light.
- **Photon.** The unit, or quantum, of electromagnetic energy (not just visible light).
- **Planet.** Natural, substantial satellite of a star. Defined by the 26th General Meeting of the International Astronomical Union in Prague 2006 as a celestial body that orbits a star; has attained hydrostatic equilibrium (round shape); and has cleared the neighbourhood of its orbital.
- **Plasma.** The fourth state of matter, after solid, liquid, and gas; a completely ionised fluid of electrons and bare nuclei, unbound and moving freely.
- **Plasma cosmology.** A description of the universe in terms of predominant electrical and magnetic fields.

- **Plenum**. A satisfied vacuum. A saturation of outward pressure.
- **Polarity**. Points in a system representing opposing characteristics; the force created by such points or poles.
- **Populant**. Astrophysical system.
- **Postulate**. To suggest that something is true.
- **Proton**. A positively charged particle, constituent of atomic nuclei.
- **Pulsar**. A small (20 – 30 km diameter), high-energy neutron star, which spins very rapidly. The densest form of visible matter.
- **Quantum**. A discrete quantity of energy, the smallest that can join or leave an energy system.
- **Quantum Hypothesis.** The suggestion by Max Planck that energy at atomic level is emitted and absorbed in discrete quantities.
- **Quantum jump (or leap).** The movement of electrons between orbital shells in Niels Bohr's explanation of the Quantum Hypothesis.
- **Quantum mechanics.** A controversial theory of particle behaviour.
- **Quantum state.** A set of properties held exclusively by a quantum for any point in space-time.
- **Quark**. Hypothetical component particle (together with antiquarks) of a *hadron*; partial electrical charge.
- **Quark confinement.** The cohesive force of quarks, zero at contact and increasing with distance, without limit.
- **Quasar**. *Quasi*-stell*ar* Object or QSO, originally thought to be highly radiant, extremely remote objects at the centre of galaxies, but later shown by Arp et al to be associated with ejecta from active galactic nuclei. Have characteristically high (intrinsic) redshift.
- **Radiation**. The means by which energy transports itself.
- **Radioactivity**. The spontaneous decay of unstable isotopes accompanied by high-energy radiation.
- **Realm**. The known or knowable universe; the Metagalaxy.
- **Redshift**. The loss of energy (increase in wavelength) in light when exposed to certain environmental factors; spectral bias towards lower frequencies.

- **Redshift anomaly**. The discrepancy between interpretation and fact in sidereal redshifts; the fact that cosmic redshifts have a multiplicity of origins.
- **Relativity.** The role of frames of reference in measurement.
- **Religion**. Cosmology that requires supernatural intervention and unquestioning faith.
- **Resolution**. The capacity of an optical instrument to separate points that are extremely small.
- **Retrodict**. Reveal the nature of an historical event by systematically tracing the sequence of events in reverse.
- **Scale warping**. A relativistic effect that distorts the results of measurements taken at great distances, macro or micro.
- **Schwarzschild radius**. The radius of a sphere into which matter must be compressed in order to form a black hole. Represented by $2GM/c^2$, where G is the gravitational constant, and M is the mass. The surface of a sphere with this radius would be the *event horizon* of a black hole, from which neither matter nor any form of radiation can escape. (See *Hawking radiation*.)
- **Seyfert galaxy**. A class of galaxy characterized by an extremely active nucleus.
- **Singularity**. In astrophysics, a point in space-time where matter becomes infinitely compressed into a volume infinitesimally small; in philosophy, a place where God divides by zero; in common sense, *nothing*!
- **Solar mass**. Astronomical unit of mass, equal to the mass of our Sun.
- **Solar system**. An organisation of plasma, planets, asteroids, comets, and other matter around the Sun.
- **Sound**. Vibrations in a medium that carry sound energy. Also the effect of these vibrations in creatures equipped with aural perception. The type of medium affects the speed of sound; sound cannot travel in a vacuum.
- **Space**. An abstract, non-material, 3-dimensional volume that accommodates the Universe and all events in it.
- **Spatial credibility factor**. Uncertainty brought about by remoteness.
- **Spectrum**. Particular distribution of electromagnetic radiation that is characteristic of the radiation itself, for example the colour spectrum of white light which displays

violet, blue, green, yellow, orange, and red bands. It speci-
fies the wavelengths present in the radiation, and how
strong they are, and properties derived from tainting as it
travels.

- **Spectroscopy.** Analysis of spectral lines in light.
- **Spin.** Rotational state; an indication of preferred attitude
in a duality set.
- **Standard candle.** The brightness of a Cepheid, white
dwarf or supernova, used as a measure of stellar brightness.
- **Standard Model of Cosmology.** Abbreviated *SCM*,
known properly as the *Lambda Cold Dark Matter Model* or
LCDMM, and referred to colloquially as BBT or *Big Bang
Theory.*
- **Star.** An incandescent celestial body or ember of such a
body.
- **Static.** In cosmology, an adjective qualifying the cosmos
such that it does not organically expand; a static Universe
is none of spreading out, becoming less dense, or growing
larger.
- **Steady state.** A set of cosmological theories attributed
mainly to Fred Hoyle, which describe an infinitely self-
sustaining, expanding Universe.
- **Strong nuclear force.** One of four fundamental forces of
nature, it is the force that binds protons and neutrons in
the nucleus of an atom; the strongest of the four (normally), it is 10^{16} times more powerful than the Weak Force.
- **Sun.** A yellow dwarf star at the focus of our Solar System.
- **Supercluster.** A conglomeration of galaxy clusters, held
together primarily by electro-magnetic forces.
- **Supernova.** An exploding or fracturing massive star; part
of the process of cosmic regeneration.
- **System.** (Energy system). An integrated arrangement of
matter comprised of quanta in equilibrium (in *coherence*).
Systems vary enormously in size and scope; the range is
probably greater than atomic nuclei to galactic super-
clusters.
- **Temperature.** A measure of thermodynamic excitement.
- **Thermodynamics.** The study of heat.
- **Tides.** Cosmic free will; the force of chaos.

- **Time**. The continuous sequence of events, flowing always from past to future.
- **Trillion**. One thousand billion; 1,000,000,000,000; 10^{12}.
- **Uncertainty principle**. Heisenberg's assertion that, at any point in time, only the position or the velocity of a particle can be known, not both.
- **Universe**. Mr, a champion bodybuilder. It's hard to be serious about this one. Probably the most common meaning of *universe* is the sum total of everything that exists.
- **Velocity**. Speed and direction.
- **Wave-particle duality**. Light sometimes displaying wave-like, and at other times particle-like characteristics.
- **Weak nuclear force**. One of four fundamental interactions in nature, second only to the Strong Force in power. It binds elementary particles together over an extremely short range. Together with EMR, it is a manifestation of Electroweak Force.
- **Weight**. The effect of *gravity* on an object with mass; it is proportional to mass. Or, how strongly an object is pulled towards the dominant center of gravity in its environment.
- **X-Stream**. A multi-disciplined journey into Life and the Universe that is open to all who want to strip away the mask of prejudice, to unlock the shackles of conditioning and indoctrination, and to start with a blank slate to determine what really lies behind the mysteries of existence.
- **Zero point**. On any scale of measurement, a complete absence of that being measured.
- **Zero point field**. The contention in quantum mechanics that at zero point (for example, of temperature or vacuum), there is still energy.

Bibliography

1. Alfvén, Hannes and Arrhenius, Gustaf. *Evolution of the Solar System* Honolulu: University Press of the Pacific, 1976.
2. Alfvén, Hannes. *Cosmic Plasma* Dordrecht: D Reidel Publishing Company, 1981.
3. Arp, Halton. *Catalogue of Discordant Redshift Associations* Montreal: Apeiron, 2003.
4. Arp, Halton. *Quasars, Redshifts, and Controversies* Berkeley: Interstellar Media, 1987.
5. Arp, Halton. *Seeing Red: Red shifts, Cosmology, and Academic Science* Montreal: Apeiron, 1998.
6. Baryshev, Yurij, and Teerikorpi, Pekka. *The Discovery of Cosmic Fractals* Singapore: World Scientific Publishing Co., 2002).
7. Behe, Michael J. *Darwin's Black Box* New York: The Free Press, 1996).
8. Böhm-Vitense, Erica. *Introduction to Stellar Astrophysics* (three volumes) Cambridge: Cambridge University Press, 1989.
9. Calder, Nigel. *Einstein's Universe* New York: Viking Press, 1979).
10. Clegg, Brian. *Infinity* New York: Carroll & Graf Publishers, 2003).
11. Close, Frank. *Particle Physics A Very Short Introduction* Oxford: Oxford University Press, 2004.
12. Copernicus, Nicolaus. *On the Revolutions of the Heavenly Spheres* New York: Prometheus Books, 1995.
13. Derbyshire, John. *Prime Obsession* Washington: Joseph Henry Press, 2003.
14. Einstein, Albert, and Infeld, Leopold. *The Evolution of Physics* New York: Simon and Schuster, 1938.
15. Einstein, Albert. *Letters to Solovine 1906 – 1955* New York: Carol Publishing Group, 1993.
16. Einstein, Albert. *Out of my Later Years* New York: Wings Books, 1996.
17. Einstein, Albert. *Relativity the Special and the General theories* New York: Three Rivers Press, 1961.
18. Euclid *The Thirteen Books of the Elements,* my copy the 1908 translation by Sir Thomas Heath New York: Dover, 1956).
19. Faraday, Michael. *The Chemical History of a Candle* New York: Crowell, 1957.
20. Faraday, Michael. *The Forces of Matter* New York: Prometheus Books, 1993.
21. Ferris, Timothy. *The Whole Shebang* New York: Touchstone, 1997.

22. Feynman, Richard P. *Surely You're Joking, Mr. Feynman!* New York: W. W. Norton & Company, 1997.

23. Feynman, Richard P. *The Character of Physical law* New York: Modern Library Edition, 1994.

24. Galilei, Galileo. *Dialogues Concerning Two New Sciences* New York: Prometheus Books, 1991.

25. Greene, Brian. *The Elegant Universe: Superstrings, Hidden Dimensions, and the Quest for the Ultimate Theory* New York: Vintage Books, New York, 2000.

26. Gribbin, John. *Stardust* Penguin Books, London, 2001.

27. Hawking, Stephen. *A Brief History of Time* London: Bantam Press, London, 1988.

28. Hawking, Stephen. *The Universe In A Nutshell* London: Bantam Press, 2001.

29. Herbert, Nick. *Quantum Reality* New York: Anchor Books, New York 1985.

30. Hey, Tony, and Walters, Patrick. *The New Quantum Universe* Cambridge: Cambridge University Press, 2003.

31. Hogan, James P. *Kicking the Sacred Cow* New York: Baen Books, 2004.

32. Hoyle, Fred, and Wickramasinghe, N. C. *Lifecloud* London: J. M. Dent & Sons Ltd., 1978.

33. Hoyle, Fred, Burbidge, Geoffrey, and Narlikar, Jayant. *A Different Approach to Cosmology* Cambridge: Cambridge University Press, 2000.

34. Jammer, Max. *Concepts of Force* Cambridge MA: Harvard University Press, 1957.

35. Kanipe, Jeff, and Webb, Dennis. *The Arp Atlas of Peculiar Galaxies—A Chronicle and Observer's Guide* Richmond VA: Willman-Bell, Inc., 2006.

36. Kepler, Johannes. *Epitome of Copernican Astronomy & Harmonies of the World* New York: Prometheus Books, 1995.

37. Kuhn, Thomas. *The Structure of Scientific Revolutions* Chicago: University of Chicago Press, 1966.

38. Kulsrud, Russel M. *Plasma Physics for Astrophysics* Princeton: Princeton University Press, 2005.

39. Kundt, Wolfgang. *Astrophysics—A New Approach* Berlin: Springer, 2001.

40. Laughlin, Robert B. *A Different Universe (Reinventing Physics From The Bottom Down)* New York: Basic Books, 2005.

41. Lerner, Eric J. *The Big Bang Never Happened* New York: Vintage Books, 1992.

42. Lerner, Eric J., and Almeida, José B. (Editors) *Proceedings of the 1st Crisis in Cosmology Conference, CCC-I* New York: AIP Conference Proceedings, Vol 822, 2006.

43. Mach, Ernst. *Space and Geometry* originally published by Chicago: The Open Court Publishing Co., 1906; my edition published by New York: Dover, 2004.

44. Marmet, Paul. *Absurdities in Modern Physics: A Solution* (Self published 1993).

45. Maxwell, James Clerk. *Matter and Motion* New York: Dover Publications, 1954.

46. Maxwell, James Clerk. *Theory of Heat* New York: Dover Publications, 2001.

47. Maxwell, James Clerk. *Treatise on Electricity and Magnetism* New York: Dover Publications, 1954.

48. Mitchell, William C. *Bye Bye Big Bang, Hello Reality* Carson City: Cosmic Sense Books, 2002.

49. Moore, Patrick. *The Unfolding Universe* London: Michael Joseph Ltd, 1982.

50. Nagel, Ernest, and Newman, James. *Gödel's Proof* New York: New York University Press, 2001.

51. Newman, James R. (editor) *The World of Mathematics* (four volumes) Redmond WA: Tempus Books, 1988.

52. Newton, Isaac. *Opticks* New York: Dover Publications, 1952.

53. Newton, Isaac. *The Principia: Mathematical Principles of Natural Philosophy*. Translated by I. Bernard Cohen and Anne Whitman Berkeley: University of California Press, 1999.

54. Peebles, P. J. E. *The Principles of Physical Cosmology* Princeton: Princeton University Press, 1993.

55. Penrose, Roger. *The Emperor's New Mind* (Oxford University Press, Oxford, 1989).

56. Penrose, Roger. *The Road to Reality, A Complete Guide to the Laws of the Universe* London: Jonathan Cape, 2004.

57. Popper, Karl. *Logic of Scientific Discovery* Abingdon: Routledge, 2002.

58. Potter, Franklin, and Jargodzki, Christopher. *Mad About Modern Physics* Harboken: John Wiley & Sons, 2005.

59. Ratcliffe, Hilton. *The Virtue of Heresy—Confessions of a Dissident Astronomer* Milton Keynes: AuthorHouse, 2007.

60. Rolfs, Claus E., and Rodney, William S. *Cauldrons in the Cosmos* Chicago: University of Chicago Press, 1988.

61. Sagan, Carl. *The Cosmic Connection* London: Papermac, 1981.

62. Schrödinger, Erwin *What is Life?* Cambridge: Cambridge University Press, 1967.

63. Scott, Donald E. *The Electric Sky* Portland: Mikamar Publishing, 2006.

64. Selleri, Franco (editor). *Open Questions in Relativistic Physics* Quebec: Apeiron, 1998.

65. Selleri, Franco (editor). *Wave-Particle Duality* New York: Plenum Press, 1992.

66. Smolin, Lee. *The Trouble with Physics: The Rise of String Theory, The Fall of a Science, and What Comes Next* New York: Houghton Mifflin, 2006.

67. *The Penguin Dictionary of Physics* – third edition.

68. Van Flandern, Tom. *Dark Matter Missing Planets & New Comets* Berkeley: North Atlantic Books, 1993.

69. Verschuur, Gerrit L. *Interstellar Matters* New York: Springer, 1989.

70. Webb, Stephen. *Measuring the Universe* Chichester: Praxis Publishing, 1999.

71. Weinberg, Steven *The Discovery of Subatomic Particles* Cambridge: Cambridge University Press, 2003.

72. Weinberg, Steven. *Dreams of a Final Theory* New York: Vintage Books, 1994.

73. Weinberg, Steven. *The First Three Minutes* New York: Basic Books, 1977.

74. Whitehead, Alfred North, and Russell, Bertrand. *Principia Mathematica* Cambridge: Cambridge University Press, 1997.

75. Woit, Peter. *Not Even Wrong: The Failure of String Theory And the Search for Unity in Physical Law* New York: Basic Books, 2006.

76. Zukav, Gary. *The Dancing Wu Li Masters* New York: Harper Collins Publishers, 2001.

77. Zukav, Gary. *The Seat of the Soul* London: Rider & Co, 1990.

Index

About the Author

Hilton Ratcliffe is a South African-born physicist, mathematician, and astronomer. He is a member of both the Astronomical Society of Southern Africa (ASSA) and the Astronomical Society of the Pacific. He is prominently opposed to the stranglehold that Big Bang Theory has on astronomical research and funding, and to this end became a founding member of the Alternative Cosmology Group (an association of some 700 leading scientists from all corners of the globe), which conducted its inaugural international conference in Portugal in 2005.

He is an active member of the organisational, scientific, and proceedings committees for the second ACG conference, which was held in the USA in September 2008. Hilton has been frequently interviewed in the press, radio, and television, and has authored a number of papers for scientific journals, books, and conferences. He writes a monthly astrophysical column for *Ndaba*, the Durban Centre newsletter of the Astronomical Society of Southern Africa, and is editor of the ACG newsletter. He serves as consulting astrophysicist on the steering committee of the Durban Space Science Centre and Planetarium, a project of the Astronomical Society of Southern Africa (Durban Centre). Hilton Ratcliffe is best known in formal science as co-discoverer, together with eminent nuclear chemist Oliver Manuel and solar physicist Michael Mozina, of the CNO nuclear fusion cycle on the surface of the Sun, nearly 70 years after it was first predicted.

In his capacity as a Fellow of the (British) Institute of Physics, he involves himself in addressing the decline in student interest in physical sciences at both high school and university level, and particularly likes to encourage the reading of books. Hilton Ratcliffe may be reached by email at hilton@hiltonratcliffe.com.

Made in the USA
Monee, IL
10 June 2026

52191945R00144